STARSTRUCK

$\Omega_k = -k(c/aH)^2$, $\Omega_m = 8\pi G\rho/3H^2$

$$S_{BH} = \frac{A}{4\hbar G}$$

Dark ages

Reionized universe

ion

$L' = L$

$L_{Planck} = ct_{Planck} \approx 1$

Reionization

smic microwave
background

$$\dot{k} = -0.5 \sum \langle r_i F_i \rangle$$

$8\pi v^2/c^3$

$$Q \sim CM^2/R \sim 4 \cdot 10^{48}$$

$\rho_v U = 8\pi v^2 U/c^3$

$n = 0, 1 \ldots$

$$S = -\sum_n P_n \ln P_n$$

N

$$P_n = A\exp(-\mu n h v)$$

$\sum_n \varepsilon_n P_n = U$

$$\cos\theta \frac{dI_v}{ds} = -\alpha_v I_v + J_v$$

$$I_v(s) = I_v(s_0) + \int_{s_0}^{s} J_v(s')ds'$$

terrestrial planets $P_v = \frac{1}{c}\int I_v \cos^2\theta \, d\Omega$

$$F = \int I_v \cos\theta \, d\Omega \, dv = \sigma_{sb} T^4$$

gigant planets

$u_v = \frac{1}{c}\int I_v d\Omega$

$$-\frac{d\Omega}{dt} = -\frac{GM^2}{R^2}\frac{dR}{dt}$$

$F = \int H_v dv$ $L = 4\pi R^2 F$

$$M_H = c/H_0 \approx 10^{28}$$

$$t_0) \approx 10^{-3}$$

$$R(t) \sim e^{H(t)t}$$

$$L \leq L_0$$

$$t_{Planck} \approx 10^{-43}$$

$$u(v,T) = v^3 \int(v/T)$$

$$u(v,T) = C_1 v^3 \exp(-C_2 v/T)$$

STARSTRUCK

A MEMOIR *of*

ASTROPHYSICS

and FINDING LIGHT

in the DARK

SARAFINA EL-BADRY NANCE

DUTTON

ion trail

nucleus

$$u(v,T) = P_v U = \frac{8\pi h v^3}{c^3(\exp(hv/kT)-1)}$$

orbit

$$2hv^3$$

DUTTON

An imprint of Penguin Random House LLC
penguinrandomhouse.com

Copyright © 2023 by Sarafina El-Badry Nance
Penguin Random House supports copyright. Copyright fuels creativity,
encourages diverse voices, promotes free speech, and creates a vibrant culture.
Thank you for buying an authorized edition of this book and for complying
with copyright laws by not reproducing, scanning, or distributing any part of
it in any form without permission. You are supporting writers and allowing
Penguin Random House to continue to publish books for every reader.

DUTTON and the D colophon are registered trademarks of
Penguin Random House LLC.

All photographs in the book are courtesy of the author.

LIBRARY OF CONGRESS CATALOGING-IN-PUBLICATION DATA
has been applied for.

ISBN 9780593186794 (hardcover)
ISBN 9780593186800 (ebook)

Printed in the United States of America
1st Printing

BOOK DESIGN BY LORIE PAGNOZZI

This book is a work of memoir. It is a true story faithfully based on the
author's best recollections of various events in her life. In some instances,
events have been compressed and dialogue approximated to match the
author's best recollection of those exchanges. Names and identifying details of
some people mentioned have been changed.

FOR MY DAD,
WHO WILL ALWAYS BE
PERCHED ON MY SHOULDER.
I LOVE YOU.

$$R_H = c/H_0 \approx 10^{28}$$

$$(t_{Plank})/R(t_0) \approx 10^{-3} \qquad R(t) \sim e^{H(t)t}$$

$$t_{Plank} \approx 10^{-43}$$

$$L \leq L_0$$

$$u(v,T) = v^3 f(v/T)$$

$$u(v,T) = C_1 v^3 \exp(-C_2 v/T)$$

$$L \frac{R^2}{R^2}$$

$$-1/3$$

ion trail

nucleus

Orbit

$$u(v,T) = \rho_v U = \frac{8\pi h v^3}{c^3 (\exp(hv/kT)-1)}$$

CONTENTS

PART I

ORIGINS

PART II

PHASES

PART III
FATES

Chapter IX

Chapter X

Chapter XI

Chapter XII

Epilogue

STARSTRUCK

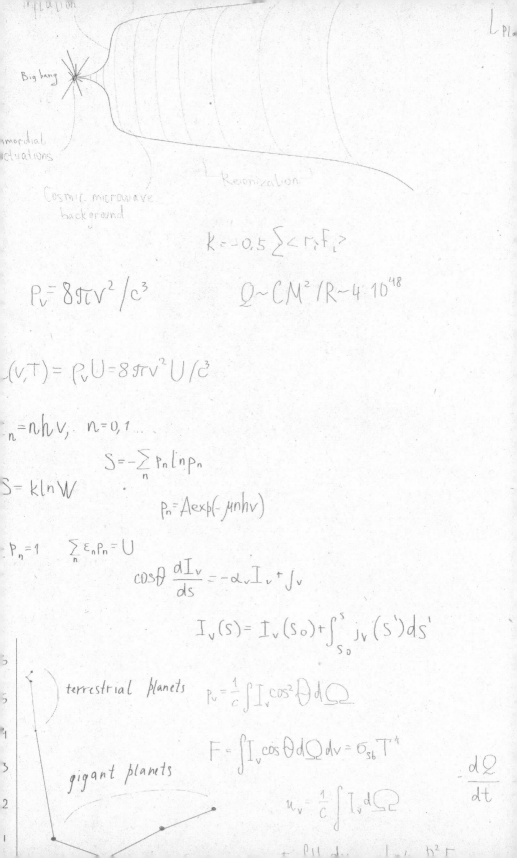

inflation

$L_{Pl...}$

Big bang

primordial
fluctuations

Reionization

Cosmic microwave
background

$$k = -0.5 \sum <r_i F_i>$$

$$P_\nu = 8\pi\nu^2/c^3 \qquad Q \sim CM^2/R \sim 4 \cdot 10^{48}$$

$$(\nu, T) = P_\nu U = 8\pi\nu^2 U/c^3$$

$$_n = n\hbar\nu, \quad n = 0, 1 \ldots$$

$$S = -\sum_n P_n \ln P_n$$

$$S = k \ln W$$

$$P_n = A\exp(-\mu n h\nu)$$

$$P_n = 1 \qquad \sum_n \varepsilon_n P_n = U$$

$$\cos\theta \frac{dI_\nu}{ds} = -\alpha_\nu I_\nu + J_\nu$$

$$I_\nu(s) = I_\nu(s_0) + \int_{s_0}^{s} j_\nu(s')ds'$$

terrestrial planets

$$P_\nu = \frac{1}{c}\int I_\nu \cos^2\theta \, d\Omega$$

$$F = \int I_\nu \cos\theta \, d\Omega \, d\nu = \sigma_{sb} T^4$$

gigant planets

$$u_\nu = \frac{1}{c}\int I_\nu \, d\Omega$$

$$\frac{d\Omega}{dt}$$

$\approx 10^{-35}$

$t_{Plank} \approx 10^{-43}$

$u(v,T) = v^3 f(v/T)$

$u(v,T) = C_1 v^3 \exp(-C_2 V$

PART I
ORIGINS

$\dfrac{dR}{dt} = L\dfrac{R^2}{R^2_r}$

$u(v,T) = \rho_v U = \dfrac{8\pi h v^3}{c^3 (exp(hv/K}$

CHAPTER I

THE MOTIONS OF CELESTIAL OBJECTS

The universe is not *is*; it *appears*. The sun and stars *appear* to rise in the east and set in the west. The moon *appears* to change shape every twenty-eight and a half days, growing into a circle and then disappearing into nothingness, setting the calendar element we call a "month." The constellations and solar system *appear* to rotate around the Earth, tracing lines of stardust across the night sky.

But this isn't the truth. Or, rather, the full truth. There is a reality—an objective, scientific, physical process—for how things move, evolve, and decay. There is an explanation for the motion of celestial bodies, a description anchored in physics. Our perspective is limited in scope: we see the universe as it appears. We grab a small corner of the picture at its ends, hands brushing along the creases as we pull, pull, pull, dragging it across the frame until that small corner brushes up against all four corners, obscuring the hidden reality behind its borders. We lay claim to it and call it Truth. We say, "This is what the universe looks like."

We're wrong.

The sun and stars only appear to rise in the east and set in the west; they do not actually traverse our skies and plow through the atmosphere each day until they rest on the opposite end of the Earth. If this were true, the Earth would be at the center of the universe. We would gaze out and see celestial bodies orbiting around, stars flaring, and planets cascading through space as we sit like kings at the universe's center, crowns alit with stars as we command our celestial empire. This is an easy picture to believe, an ego-centered approach that's survived for thousands of years, mirroring our lived experiences. We are the protagonists of our own stories, the heroes in our fairy tales.

But it's not the truth.

The Earth is not the center of the universe. It is a small, fragile world floating upon the waves of spacetime amongst a sea of worlds. It orbits an average-size star alongside seven other planets, making up the local cosmic neighborhood we call the "solar system." What makes the Earth remarkable is not its position in the universe—many Earth-size planets orbit their stars close enough for liquid water to exist—but what inhabits it. It is not just the Earth's liquid blue oceans but its vast desert plains too; the oxygen-rich atmosphere and solid molten core; the millions of species and the complex, diverse ecological systems—all of which combine in an exquisite, unique dance to make the Earth special.

It is the stuff that makes Earth ripe for life.

The Earth is the third planet from the sun. But because the Earth *spins*, rotating around its slightly tilted axis, the sun appears to rise and fall, ushering in day and night. Every twenty-four hours the Earth rotates around its axis, plunging half of the planet, the side facing away from the sun, into darkness and the other half of the planet, the side facing toward the sun, into sunlight.

This is the complete picture, the invisible explanation that we cannot actually feel because we are stuck to the Earth by gravity. As the Earth flies through space, so do we.

Over the course of a night, the other planets in our solar system follow the same line as the sun across our skies, rising in the east and setting in the west. But over the course of a year, they appear to carve out their own paths, wandering across spacetime seemingly of their own accord and drifting against the crowd of stars from west to east. Contrary to our intuition, this is not because these planets are traveling on different trajectories through space but because they are orbiting the sun more slowly than the Earth is. The apparent difference in speeds between the Earth's orbit and the orbits of these planets creates the illusion of "retrograde" motion, as they appear to travel backward across our sky.

The Earth drags the moon in tow as it completes its orbit around the sun. To us, the moon appears to change shape, growing into a perfect circle and then fading into nothing over a twenty-eight-day cycle. But the moon does not swell and shrink. In fact, the moon itself orbits around the Earth just as the Earth orbits around the sun. When the moon is on the opposite side of the Earth with respect to the sun, the sun's rays perfectly hit the lunar surface, illuminating the side facing us and making it a full moon. On the other hand, when the moon is on the side of the Earth nearest the sun, the sun's rays are blocked, obscuring the moon and plunging it into darkness. In this configuration, we see a "new moon," which looks like a moon just out of light's reach. But in reality, we're just not seeing the other side of the moon, the one hit by the sun's rays.

This is contrary to what we might think as we rest our feet atop blades of grass and gaze up at the sky. We convince ourselves that what we see must be true. Our eyes take in the motions of the stars and the apparent reality of the planets painted across the canvas of the night sky and we think, *This is it.*

But our eyes can't possibly reveal the truth: that we're on a rock hurtling through space at thirty kilometers per second. And that our cosmic neighborhood, our solar system, is in fact orbiting the center of our galaxy, a supermassive black hole called Sagittarius A* (the asterisk conveying the excitement of discovery, since atomic physics designates excited atomic states with an asterisk). And this galaxy, our home, is just one of billions. At some point, numbers cease to mean anything as our brains grapple with the incomprehensible scale of the universe.

It's difficult to fathom what we can't see or touch, taste, or smell. But we force ourselves to continuously push beyond what we see right in front of us for a chance to glimpse the unknown. We pen poems and solve math problems, paint in watercolor and write code, read books upon books and peer through telescopes to learn more about our home, the universe.

These pictures are still incomplete. We view the world and our place within it through the lens of our lived experiences, our own struggles and dreams. But it is in this messy place of complexity that we have a chance of learning something fundamental and true.

We just have to be brave enough to question our perceptions. Somewhere underneath our biases and limited perspective is the entire universe, waiting to be discovered.

We just might not understand it yet.

I've spent a lot of my life in cars. During the school year, from kindergarten through senior year of high school, I drove forty-five minutes to and from school each day. The drive could be as long as an hour and a half if there was bad traffic clogging up Highway 360. When you add in the trek to tennis practice (on one end of Austin) and tutoring (on the other end), I probably spent three total hours every day in the seat of a car.

I didn't hate the drives, partially because I could catch up on missed sleep and partially because of *StarDate*.

Every day, at exactly 6:57 A.M. and 4:57 P.M., Sandy Wood's voice would float through the speakers in our silver Volkswagen while she hosted a two-minute astronomy show. In order to listen, I had negotiated a hard-fought deal with my parents: I agreed to listen to their choice of classical music and five-o'clock news if they let me listen to *StarDate*. It took less convincing than I expected—my mom hid an odd smile of pride when I proposed the deal—but I refused to think too hard about it, and with a quick *thank you* was able to spend four minutes per day learning about the stars.

In first grade, mornings begin at 6 A.M. By 6:15, I've packed my backpack and furiously checked that I had all my homework before my mom and I slide into the old family Volkswagen. By the time we leave the driveway, the sun is still below the horizon and darkness cloaks the Texas landscape. I sit rigidly in the backseat, eyes scanning across lines of homework problems to double-check my work, while we drive from the outer suburbs to the other side of Austin, where my school, St. Andrew's, is. The houses grow bigger and bigger over the course of the drive, rising with the sun as we pass the gated neighborhoods of Westlake and estates of Tarrytown.

My mom's scent washes over me as we drive, lemongrass and pumice stone laced with cigarette smoke. She refuses to smoke in the car or the house, claiming she wants to shield me, but it has long ago seeped into her clothes and the crevices of her skin. I scrunch my nose when a wave reaches me.

Ethereal music soars through the speakers, signaling the beginning of a new episode of *StarDate*. "Mom, can you turn it up?" She turns the dial and Sandy's voice grows louder. I relax into my seat and count the trees whipping by as we round the corner onto Bee Cave Road. They look like broccoli, the dark-green leaves of cedar displayed like chunks of great big florets.

"Today on *StarDate*: Venus, Earth's sister, is visible just above the moon at twilight. It will look like a bright star, but unlike other stars strewn across the night sky, it won't flicker." I let her voice, imbued with the magic of Venus and the cosmos, lull me into a semi-meditative state as we get closer to school. The sky brightens over the course of the drive, transitioning from black to orange to yellow, and when it finally hits pale notes of blue, we are pulling into the long driveway marking the entrance to St. Andrew's.

St. Andrew's is what you get when you combine white liberal Austin with the Episcopalian church. The elementary school is auspicious; a large cream-white slab of marble marks the entrance, "Crusaders" engraved in bold. It was unclear what St. Andrew's meant by calling ten-year-olds crusaders. And what were they crusading for: the crucifixion of their enemies? Nonbelievers?

Beside the slab is a flagpole, flying a triumvirate of flags: the United States, Texas, and the St. Andrew's flag. The Texas flag is larger than the other two, but just barely: our school flag makes up for its size with fat white stark lines slashed across a pale-blue background, domineering in its frankness. The American flag, on the other hand, is almost perfunctory, flying amongst the others yet never fully visible.

✳

"Can we look at Venus tonight? Did you hear Ms. Wood say it'll be visible before sunset?" I ask my mom plaintively as we pull up behind a forest-green Mercedes. She smiles a bit, the sun emerging from clouds as she tucks a stray strand of chemically straightened brown hair behind her ear. No matter how many times she has tried straightening it, going to expensive salons and spraying bottles called "sleek" and "soft," her hair stays frizzy, the Egyptian coils refusing to fully succumb to the beauty ideals of the West. When I see her touch her hair, I subconsciously reach for mine, reassuring myself that my curls are temporarily tamed before entering the building.

"Maybe, but you have tennis at five and we need to go through your spelling homework to review for your test on Friday."

"But that's in three days!" I protest. "I have time!"

"We don't want to disappoint your dad, do we?" Indignant, but realizing the futility of fighting back, I suck in a quick breath. She twists to look at me, the weight of expectation sharpening her gaze.

"No," I mutter, "we don't." I grab my backpack and leave the car without saying goodbye.

I stick my backpack into one of the light-blue cubbies lining the hallway, painted to match the school colors, and take a seat at my desk. My name is written on a sheet of paper folded up like a tent at the edge of the table, scrawled in orange and yellow highlighter. On the other side of the sheet, the one facing me that people couldn't see, I'd inscribed my name in Arabic: three seemingly distinct groups of lines and curves, four dots and a long *elif.* I love writing my name in Arabic, love feeling the beauty of the curvature of the letters as I scratch them onto the page. The Arabic alphabet isn't quite as beautiful as the alphabet of hieroglyphics we have hanging in our living room at home, but it's far prettier than the English one, made up of coarse lines and brutish scratches. I straighten the name-tent, bending it slightly to hide the Arabic from full view.

My mom taught me how to write my name in Arabic using my toy chalkboard. When she took the pink chalk—I wanted blue—and guided my hand through the shapes, she said that when she was a child growing up in Egypt, her mom taught her how to write too. That was before bombs fell from the sky, before her grandparents constructed a shield of sandbags to protect them from the air raids.

Before she left Egypt.

My grandfather Mohamed was a director of demography for the United Nations. His job took my grandmother, my mom, and her three brothers all over the world. First was India—my mom's favorite, she says, the place that felt closest to home. Then was Switzerland, which was beautiful and cold and where she learned to hate her Egyptian curls. Finally, they relocated to the headquarters of the United Nations: New York.

My mom loves New York. She pronounces it with the slightest hint of a British accent: *New Yohk.* I think she resents moving to Texas for graduate school. New York has culture, community, *life.* She doesn't stand out in New York. No PhD will ever be worth leaving that behind, even if graduate school brought my parents together. Even if she now gets to share the title of *Dr. El-Badry* with her father.

Texas, she says, will never be her home.

I wonder if it will ever be mine.

The first-grade desks are organized with purpose: three groups of four desks each, distributed like a fractal in the center of the room. On the floor is a rainbow-colored carpet with the periodic table of elements printed on top, hydrogen on the side closest to the chalkboard and radon at the back by the sink and art supplies. I sit on top of oxygen, my second most favorite element next to neon, which I'm told glows like the sun.

I count the homework pages in my blue homework folder to triple-check that I haven't left anything at home. My heart thumps as I get to the last page, and I scan it twice to make sure it's fully completed with no obvious errors. I finish checking the last problem right as the bell rings and Mrs. Fitzgerald asks us to stand to pledge allegiance to the flag of the United States. Everyone crosses their heart with their right hand, but my mom once told me that I don't have to, so I stand tall, my hands down by my pockets. Mrs. Fitzgerald looks at me reprovingly, and I immediately bring my hand back up.

A pair of girls, their blond hair glittering in the sunlight, nudge each other when Mrs. Fitzgerald calls for attendance and reaches my name, her Texas drawl drawing out *Saraaafeeena*. Once she gets to *Zubin, Zach*, we line up at the front of the classroom. I make sure to get near the front, waiting for Mrs. Fitzgerald's approving smile before leaving the room.

The floors of St. Andrew's are tiled with square-foot white and gray squares. I make sure to keep my feet inside the squares, counting one after the other as we make our way to chapel. One of the blond girls laughs at something her friend says and pushes her lightly out of line. Mrs. Fitzgerald stops us to reprimand the two girls, pulling them to the front of the formation to keep an eye on them.

That's what happens when you don't follow the rules, I think to myself, smirking. I remind myself to stay perfectly in line and follow their lead into the chapel.

Every day at St. Andrew's begins with a forty-five-minute chapel ceremony, unless it is Ash Wednesday or Good Friday or Ascension Day or the Day of Pentecost or any major church event. On those days—the ones I dreaded—morning classes were canceled and the chosen few dressed in heavy religious garb designed to usher us through prayer and reflection and repent our many sins.

There were many to repent for, we were told! What we ate for breakfast, how we spoke to our teachers, the homework we didn't complete and the class assignments we did poorly, the music we listened to, and the original sin of just being alive. As we prayed through the cycle of guilt, I'd kneel on the pews in silence. I'd close my eyes, zip my lips, and focus on my breathing—in, out, in, out—as their voices echoed off the sepulchral room.

The Episcopalian chapel was designed with the austerity of Romanesque architecture, but something had gotten twisted along the way. Red velvet carpet decorated the floor, and each row boasted velvet kneelers and newly finished wooden pews. Hymnals and prayer books were replaced annually, ensuring that the golden lettering never faded from the covers. They glittered from the backs of the pews as you sat down and faced the altar.

Although the altar was a permanent fixture, intricate and striking, it was the seasonal decorations marking each religious holiday that made the chancel beautiful. Some days candles and white lilies dotted the altar, and other days heavy metal crosses were laid across the prie-dieu. An organ the size of the art classroom marked the narthex at the entrance, gleaming with its half-a-million-dollar price tag (a tax-deductible donation from a parent and patron of the church). Two heavily adorned doors stood tall

at the exit; at one point they might have been beautiful, but to me
they just looked garish.

The organ plays as we file into the chapel and take our place
in the pews, feeling the wood dig into our backsides. The minister stands at the altar waiting to begin, smiling at different classes
of students and flipping through the Bible. He looks more football
player than pastor: He is young, perhaps in his late twenties or
early thirties, six-foot-two with dirty blond hair and empty blue
eyes. His charcoal-black robes fall along the rigid lines of his body,
and though they do not make him appear closer to God, they do
make him look closer to power.

His chest fills and his voice booms out across the chapel.

"Good morning, Crusaders. Let us turn to page 682 and join
together in singing 'Let My People Go.'" Excited murmuring fills
the room as students and teachers flip open their hymnals. The
song is a crowd favorite, and the minister knows it, grinning to
himself in benevolence as he looks down from the altar.

> *When Israel was in Egypt's land,*
> *Let my people go!*
> *Oppressed so hard they could not stand,*
> *Let my people go!*
> *Go down, Moses,*
> *Way down in Egypt's land;*
> *Tell old Pharaoh*
> *To let my people go!*

My tongue furls with the tang of bitter anger as I clamp my book
of hymns shut. I hate this song. The book stays closed in my hands,

and another girl with curly hair, Nadia, who is standing beside me, looks at me questioningly. I give the tiniest shake of my head and mouth *Listen to the words.* She looks back down at her hymnal, reading through the verses silently, and looks back at me, struck.

"This is terrible," she whispers. I nod, the sick feeling in my stomach releasing a bit with her acknowledgment. She closes her book, and we stand side by side in silence, the chorus of "Let My People Go" rising to a crescendo around us.

I think about my family in Cairo, my aunts yelling "khalas!" to my cousins as they ran around the house, the sweetness of freshly cooked molokhia wafting through the air as lamb sizzled on skewers, kofta lining our dinner plates. I think about how last time we visited, my aunt Susu cooked me a birthday cake in the shape of my favorite stuffed animal, Stewy Bear, sky-blue puckered icing outlining his imaginary tie atop an extraordinary meringue cake, made specially for me. I remember Khan Khalili, the outdoor bazaar and spice market in Old Cairo where the sweet smoke of hookahs created clouds of dreams, richly dyed colored rugs lined with geometric patterns and tassels at their ends spread out on the floors, rooms of gold, geodesic drums and hand-crafted sitars, plush handsewn camels and tin replicas of pharaohs, freshly ground spices of cardamom and coriander and saffron, and gallabiyas dangling from clotheslines skirting the streets.

The air in Cairo feels different. It's not just the pollution, imbuing the air with fumes and oil, nor is it the red clay dust of the Sahara permeating the air, but the effervescent and boisterous joy bursting from the seams of its people. It is the love for the place, a city as old as time, far older than those in the West can fathom.

In Cairo, there is no such thing as loneliness. There is the warmth of your family and the invisible embrace of the land, and together their magic flows through the Nile and the pyramids, the scents of the markets and the dust of the desert, all the way to your body, down your esophagus, past your lung cavity and into your heart, where that magic sits and lights up your world.

The words of the hymn ring violently in my ears. I look around at the chapel, out at a sea of varying shades of white. The two blond girls sing in harmony together at the end of the pew, their warm, beautiful voices in tandem raised for the final crescendo as they shout sweetly, *"Let my people go!"*

I taste bile and remain silent, and the minister finally raises his hand.

"Please, sit." We obediently sit. My fingers tingle, and I clamp them down in my lap to hide the shaking.

"Today, I am going to talk about the beginning of it all." He gestures widely, excitement sharpening his gaze and spittle forming at the edges of his mouth. He continues as he strides from one side of the chancel to the other. "Of Adam, and Eve, and original sin"—I roll my eyes—"and the universe." I prepare to fully tune out, but at the word "universe," I pause.

"I am going to read a passage to you from the Book of Genesis, chapters one and two."

He begins, *"In the beginning God created the heavens and the earth. The earth was without form, and void; and darkness was on the face of the deep. And the Spirit of God was hovering over the face of the waters. Then God said, 'Let there be light'; and there was light. And God saw the light, that it was good; and God divided the light from the darkness. God called the light Day, and the darkness He called Night. So the evening and the morning were the first day."*

I look around the room, expecting to see eyes glazed over and heads lolling, as we typically do in chapel, but to my surprise the students look on, transfixed. I follow their lead and look back up at the altar.

"Then God said, 'Let there be lights in the firmament of the heavens to divide the day from the night; and let them be for signs and seasons, and for days and years; and let them be for lights in the firmament of the heavens to give light on the earth'; and it was so. Then God made two great lights: the greater light to rule the day, and the lesser light to rule the night. He made the stars also. God set them in the firmament of the heavens to give light on the earth, and to rule over the day and over the night, and to divide the light from the darkness."

The minister pauses and closes his Bible. The golden letters on the front once again shine upward, their gleam lighting his face.

"Do you see?" He's on the verge of shouting, passion lacing his voice with a deep boom. Heads nod synchronously.

"God, our Heavenly Father, created the heavens and the Earth, light and darkness. From his bounty came the grass and the oceans, our food. He created the stars in the night sky and the planets in our solar system. He created your dog." He points at someone in the front row, then turns to point at someone else: "And your cat." Swiveling once more, pointing at me: "He created you"—and, finally, pointing at himself—"and me.

"Our God in his glory created our universe, and we get to experience its beauty every single moment in every breath." His eyes shine as he looks out upon us. His faith is beautiful, and I find myself leaning forward, drawn to him. If God is responsible for the world, then everything will be okay—right?

"But the peace God gifted the Garden of Eden was stolen by the avarice of humankind. The devil took the form of a snake and gave unto Eve a bite of the forbidden apple. She took the apple in her hands, against the word of God, and ate it, and then offered it unto Adam, who also took a bite. And so they were kicked out of the garden, banished for their sin, and we must spend the rest of our lives repenting for her mistake." The minister looks sad, disappointed in Adam and Eve's betrayal. His knuckles, clutching the Bible, are white.

"We are, all of us, tainted. This is the original sin we must expunge from our bodies and our minds. We were born into sin. And so we must repent, to give thanks unto God for his glory."

I feel the flame of guilt lick my lungs, my body already responding to his call for self-reproach and shame. But at the same time, I am puzzled: How am I responsible for Adam and Eve? How can I bear the burden of their mistake? How is my *birth*—a choice I didn't even make—a sin? My sin?

"Just as Moses struck his staff into the water of the Red Sea, transfiguring the water into blood, God punishes those who do not repent. The blood of the Egyptians has permeated the land and waters of Egypt. The evidence is still there, even today, in the Red Sea!"

I pause. The Red Sea isn't *red*. I visited last year, when we went to Sharm El-Sheikh for a weekend to see our family by the sea. The Red Sea had fish all colors of the rainbow, unique flora and deadly pink coral, but it was as blue as any body of water I've seen.

The minister ends chapel with the Lord's Prayer and dismisses us with a perfunctory wave. The rest of my classmates file out of the room, but I stay behind, pretending to straighten the prayer books in the shelves of the pew. As the chapel empties and the minister begins to walk toward the exit, I step in front of him.

"Excuse me sir, I have a question about the sermon today."

He smiles down his nose at me. "Of course. How can I help you?"

"Well, sir, you see, I'm half-Egyptian. I go to Egypt every year to visit my family." I pause at the look on his face as I see his eyes harden, his lips straightening into the line of a snake's tail as it poises frozen, ready to strike. He motions for me to continue.

"I've been to the Red Sea, and you said it was red from the blood of the Egyptians but . . . it's not red. I just wanted to let you know. It's blue. It's a normal sea." I am rambling by the end, my words tripping over themselves. I point to the light blue of the Crusader flag in the back of the room. The minister raises an eyebrow and, to my great surprise, pats me gently on the back.

"I know it might seem that way to you, but you must not have been looking properly. The Red Sea turned red," he says bluntly. "The Bible is very clear—I can show you the excerpt if you like." The ends of his lips curve up in what he must think is an indulgent smile, but there's no light in his eyes.

"I . . . uh . . . but I saw it?" I protest in a small voice, looking up at him. The organ has stopped playing, plunging the chapel into an uneasy quiet. The doors slam shut as the last student exits the room, and with her exit the last vestiges of air seem to escape.

"Well, you're wrong." His words are scathing with finality. "Your eyes deceived you." With that he brushes past me and continues toward the exit, until he pauses one last time in front of the door. He looks back over his shoulder at me, finding me standing at the edge of a pew.

"You should return to class. You're late, and we do not tolerate tardiness at this school." He turns again, robes whipping behind him as he strides out the door. I collect myself and run back to Mrs. Fitzgerald, perfectly stepping within the square tiles, all the way back to the room of white and the math packet I'd left on my desk.

By the time I meet my parents outside for after-school pickup, I've had the day to calm down. I read *Harry Potter and the Sorcerer's Stone* during recess, carrying it around the playground like an emergency blanket. When Nadia asks me to play an Indian in Cowboys versus Indians, I join. We win, to our surprise—at St. Andrew's, the Cowboys always win. (I never get to be a Cowboy.)

The line of cars at pickup is already all the way out to the street. At St. Andrew's, money speaks loudly, but New Money speaks louder. Glittering Land Rovers pull up behind new Mercedes and BMWs as moms with freshly highlighted blond ponytails hop out of the front seats, their designer purses knocking into their Pilates-toned legs as they help their child into the back. Most cars have televisions installed into the back of the headrests, and though I'd begged my parents to get a new car with a TV, my parents declined, saying our five-year-old Volkswagen works just fine, thank you.

We are well-off enough to afford St. Andrew's—my dad is an energy economist who travels all over the world to help people buy and sell energy. My mom is a demographer and research associate affiliated with the University of Texas, and with the help of my babysitter, Maria, my parents shuttle me to and from school, sports practices, and playdates.

But the wealth that schools like St. Andrew's require—the wealth to fit into status brackets with families like the Dells (of Dell computers) or the Bushes (of Presidents Bush), the Crenshaws (Ben Crenshaw, a pro golfer) or the McNairs (the billionaire and owner of the Houston Texans)—entirely eludes us. The social strata that divide the more privileged areas of the United States filter through the classes at St. Andrew's, even in elementary school. Kids with hundred-dollar backpacks and perfectly organized organic, gluten-free, non-GMO lunches go to church and cotillion together. They spend Christmas break skiing black diamonds in Telluride and summers relaxing on private beaches in the Bahamas. The rest of us—still enormously privileged, but with no "fuck-you money," as my dad calls it, watch on from the sidelines, aware enough to know that we don't fit in, but not enough to fully understand why.

I slide into the backseat of our car and, to my surprise, see both of my parents in the front seats. "Dad! You're here!" He turns around in his seat, seatbelt digging into his neck as he grins. The top of his hair brushes the roof of the car, all six-foot-two of him barely fitting into the hatchback. My chest swells at the sight of him, a giddy warmth spreading through my limbs. He is my sun, radiating light to his surroundings, through my body and electrifying my heart.

"I flew back early so I could watch your tennis tournament this weekend! You ready to champ up or what?" he asks, referring to Champs, the middle-tier division of Texas junior tennis that I'm trying to qualify for. Anxiety pools in my stomach, replacing the warmth I'd just felt, but I keep my face cool, intent on keeping my fear to myself.

"I'm ready," I say. He nods approvingly.

"And how's school? Did you sort out that mistake on the spelling test from last week?" I open my blue folder to show him the lines and lines of spelling words I'd written for practice. I'd had to repeat "anomaly" fifteen times—I kept mixing up the positions of the *a* and the *o*. My mom's cursive was scratched across the top of the page, *Spelling Practice: 03.02.98*. He scans the page, taking note of the difficult words, and then hands it back to me.

"Good work, honey. We'll go through it a few more times before Friday just to make sure you're in good shape. And what did we learn from last week's exam?" He cocks his head, letting the question hang.

The previous week, my dad had been out of town, and I called him in panicked tears, entirely distraught that I'd misspelled a word on the test. My mom had tried to comfort me, telling me I still did well and she loved me regardless, but it was my dad who sat on the phone, hundreds of miles away, and listened to me sob about being a failure. On that call, we'd made a pact to do better next time. The new plan consisted of the two of us practicing spelling together for an hour or two after school each day, to drill the words so far into my brain that it would be like they'd always been etched there.

"That practice makes perfect, and I can succeed as long as I try hard enough." My voice breaks a bit on "enough," but I don't think either parent notices.

Dad resumes looking at the dash, and it is only then that I realize my mom is gripping the wheel so tightly that her hands are shaking. I wonder what I've walked into and how I missed the anger charging the air with tiny electric zaps. Although we're stopped at a red light, the Eye of Horus inscribed on the gold bangles hanging from her right wrist to give her protection jiggles up and down.

"Is everything okay?" I should have known from the second I opened the car door that something was wrong. When my parents are together, there always is.

This time, my mom answers.

"I don't know, Peter, is it?" My mom looks at my dad, daggers sharpening in her gaze. In return, he sucks in a breath through gritted teeth.

He pleads quietly. "Samia, let's not do this right now. Wait till we get home, like the therapist suggested, when Sarafina's at practice." She exhales a pained half laugh and looks back at the road, foot on the accelerator.

"Sure." Irony coats her voice in tiny thorns. "Put it to the side, as always. As if that's ever going to fix anything. I wrote down the exact words that you said, Peter. It's all here." The car jolts to the side as she pulls into a gas station and comes to an abrupt stop. The keys are still dangling from the ignition, the metal scarab on her key chain rocking back and forth as she gets up and slams the car door.

Outside, she shuffles through her purse, searching for a pack of cigarettes. My dad sits with me in the car in silence for a moment, staring at nothing. He turns back to me, plasters on a smile, and says, "It's going to be okay."

I feel my head nodding automatically, and he, too, gets up and exits the car. From outside, my mom's voice rises and cuts like a knife, sliding through the one-hundred-pound steel door and into the car, all the way to where I sit, stuck in the middle seat.

I reach into my backpack for *Harry Potter* and bend my neck low enough that I won't be able to look out the window, even if I try. I've already read *Harry Potter* so many times that I earmarked pages I found especially funny. I flip through to find the familiar creases: the one where Molly Weasley can't figure out whether she is talking to twin Fred or George, the one where Hermione lights into Harry and Ron for dragging her to the forbidden third floor. After reading enough passages, the scene outside recedes, and I even let out a giggle when they find Fluffy.

The only indication of passing time is the radio-show rotation—*All Things Considered* has just come on, and Ted Clark is reporting that the UN is threatening Iraq. When I finally peek out of the window, I see a gleaming white light next to a crescent moon. Venus, I remember.

She is remarkably bright, and as I'd learned from Sandy Wood that morning, solid in her shine. Stars, Sandy taught me, twinkle, but planets do not, thanks to their proximity to our atmosphere.

The longer I look up at Venus, the fainter my parents' voices sound; her shine dulling even my mom's loudest yells.

By the time my parents reenter the car, Harry, Ron, and Hermione have served detention in the Forbidden Forest, Hagrid has won (and lost) a dragon, NPR host Linda Wertheimer is talking about a man named Rupert Murdoch, and Venus has assumed her perch beside the moon. The air is lighter, and when they smile back at me, I know that there is a lull in the storm. The storm will return, as it always does, but the lulls are precious and rare, and I remind myself to savor the break.

Mom drives back onto the highway, and I take advantage of the newfound peace.

"Today we learned that God created the Earth and the sun and the planets. And aren't there, like, a lot of planets?" Both parents straighten, and I see them eyeing each other before my mom responds.

"There are," my mom says carefully.

"But if God created all of those planets . . . Wait, don't planets have stars too?" The scale of things is beginning to overwhelm me, derailing my original question.

My dad chimes in this time. "Many planets do have stars, yes. Some stars have no planets but some stars, like the sun, have many planets."

"But . . . there are a lot of stars, right?" Both heads nod.

"Like, hundreds?" My eyes widen.

"Try billions," Dad says. I stare at him, waiting for that to make sense. He obliges my silence. "That's nine zeroes." At this my jaw falls open.

"God created *all* of them? But if he created the universe, then who created God?" My mom smirks and looks at my dad, who is smiling before responding.

"That's a good question, honey. What do you think?"

I shrug, nonplussed. "I don't know."

My mom, without missing a beat, responds, "We don't say 'I don't know' in this house, Sarafina. What do we say instead?" She grips the steering wheel to pivot and look me full in the face.

I sigh, feeling a little defeated. "I don't know but I'll figure it out," I say, reciting the household rule back to her.

Mollified, she nods and swivels toward the front. My dad looks back at me through the side mirror and winks, rolling his eyes a little bit.

I continue, "But if someone created God, then who created *that* God? And the God after that?" I'm imagining Russian nesting dolls, one inside the other, going on and on and on until . . . what? Infinity? That there is no end is totally bewildering. I shake my head to try to understand as we finally turn into our driveway and wait for the garage door to rise.

"Another good question. One worth thinking about, don't you think?" my mom asks, pulling into the garage and shifting into park. I tear my gaze away from the door and look at them, stunned.

"You mean we don't know?"

"We don't know *yet*," my mom corrects me, but still the enormity of the acknowledgment hits me with full force. I fall back into my seat, my head swimming.

"But there's so much out there. The universe is so big, and we don't know where it came from?" I ask, disbelief lacing my words. For some reason, the thought hasn't occurred to any of us to continue the discussion inside the house, so we sit there, sweating in the leather seats amongst piled-up boxes and old jumper cables, tool kits and jars of old paint, our bicycles and tennis rackets and soccer balls, talking about the scale of the universe.

"Here's the thing, Sarafina." My dad turns around to fully face me, his blue eyes piercing through the thick lenses resting on the bridge of his nose, golden hair sweeping across his wide forehead. "St. Andrew's is a great school. But the things they teach you in chapel come from the Bible, right? And the Bible was written by humans centuries and centuries ago. And humans aren't perfect— we all have flaws, we all make mistakes. Often, we don't fully understand things. This might be the way we understood the universe a thousand years ago, but science has evolved. We know so much more now. And there's still room to explore."

I nod thoughtfully, chewing on his words.

"Can I explore it?" I ask both of them as we unbuckle our seatbelts, and my dad props open his door.

"Of course, honey. If it's something you're interested in," my dad says, and I'm excitedly envisioning the Earth and our sun, then the planets in the solar system, then *more* stars and planets, and more and more and more until I am giddy at the sheer vastness of it all.

As we make our way inside, I dump my backpack onto the kitchen table and continue to the backyard. The sun had set hours ago, plunging our side of the world into darkness. Tennis practice was long forgotten, thanks to the fight at the gas station. My dad strides after me, following our unspoken agreement that he will grab the binoculars from the counter and shut the door behind us, sealing off the backyard from my mom.

Together, we sit side by side on a couple of pool chairs. I draw my knees close into my chest and tilt my head back, hoping that the farther back I go, the more stars I will see. I try once again to wrap my mind around the scale of the universe; I picture planets orbiting stars and galaxies directing the orbits of billions of these glowing orbs of fire, their lights glittering from millions of light-years away all the way across space and time to me and my dad, sitting right here on Earth, looking up.

We sit there as the sky turns and the moon rises, until lights flicker on and my mom calls us inside. My dad begrudgingly obliges her, but I stay just like that, sitting under the canopy of stars, alone in the backyard, staring out into the universe.

CHAPTER II

OUR COSMIC NEIGHBORHOOD

When we think about our place in the world, we might define our address as this, with some variation on the details depending on where you live:

> *Number, Street*
> *City, State, Zip code, Country*

Often, we assign location relative to other objects, and say *Third house on the left down on Wellesley Drive, next to the white house with a chimney and picket fence.*

But when we think about our place in the *universe*, the scale is suddenly much larger and our sense of location far broader. It is nearly impossible to fully grasp everything that is our universe, let alone gain a sense of our place within it: the billions upon billions of planets, orbiting single- and double- and perhaps even triple-star systems, all housed within galaxies holding hundreds of billions of stars, nestled between thousands of other galaxies . . .

The scale is overwhelming. But we still try. It is human nature to derive meaning from our place within the universe. And so we try to expand our perspective and widen our scope to include *all* of *everything*.

Part of that understanding is finding a way to contextualize distance. Distances are measured differently on Earth than they are in space: On Earth we might say something is a few kilometers (or miles) away. But the moon, our closest celestial object, is a whopping 238,000 miles from our planet. Our closest star, the gigantic ball of fire and gas we call our sun, is a prodigious 93 million miles away. If we were to define cosmic distances in terms of kilometers or miles, our Earth-based units, the numbers would quickly approach something so large that they would be nonsensical.

Light is a useful measuring stick. Photons, or little packets of light, stick to the fabric of spacetime, traveling at a constant speed through the cosmos at a mind-bending 3×10^8 meters per second. Light is thus dubbed "the cosmic speed limit," and its consistency allows astronomers to do something clever: by using the speed of light, we can measure how far away distant objects are from us.

For example, it takes light 1.3 seconds to travel the distance from the moon to the Earth, so we say the moon is 1.3 *light-seconds* away. Similarly, it takes light eight minutes to travel from the sun to the Earth, so the sun is eight *light-minutes* away from us. Light-years are just one unit of distance measurement that astronomers use—a little over three light-years makes one parsec.

Imagine boarding a state-of-the-art spaceship, one decked out in silver livery with a sleek, aerodynamic finish allowing the spaceship to do the impossible: travel at the speed of light. You nestle yourself into your seat and say hello to the crew members, sipping a beverage adapted to withstand zero g, and zoom off our planet into the depths of space. On your way out of the solar system, you zip by Mars and wave hello (and then goodbye, for you are traveling quite fast) to the rovers crawling on its surface. You pass by the Great Red Spot, a violent storm that has raged for centuries on Jupiter. You dance in between Saturn's rings of ice and rock, bearing witness to the procession of geometric patterns, and then make your way to the ice giant we call Uranus. You fly even farther beyond Uranus's rings, all the way to the last planet in our solar system, Neptune, and then you are beyond.

You are pleasantly surprised to learn that the solar system doesn't end there; you continue to fly on to the dwarf planets, the most famous of which you recall is Pluto. And then, finally, you are weaving in between icy planetesimals marking the very outer reaches of our solar system, the Oort cloud.

Completing this entire journey at the speed of light, flying all the way from our planet to the far reaches of the solar system, has taken over one Earth year. You have celebrated one birthday, perhaps marking the occasion on a calendar you and your crew hang in the mess hall, yet because you're traveling at the speed of light, you feel no time passing at all. Still, for the sake of imagination, let's assume our timescales are measured from the perspective of someone on Earth.

If you were to continue to our nearest star, Proxima Centauri, at the speed of light, you would need to fly for another *three years.*

Beyond Proxima Centauri are 100 billion stars—and you are still within our home galaxy, the Milky Way.

Fly for another twenty thousand years, and you would reach the core of the Milky Way. A supermassive black hole, Sagittarius A*, gorges itself at the galactic center on nearby accreting material, augmenting the black hole's already monstrous 4 to 5 million solar masses.

Twenty thousand light-years, and you have not even left our home galaxy.

Go another 10 million light-years and you'll see our neighborhood of galaxies, the Local Group. The Milky Way and its closest sibling galaxy, Andromeda, the Local Group's two largest galaxies, circle each other in a galactic dance that will at some point, billions of years from now, end in their collision. Attached to each of these behemoths are smaller satellite galaxies, gravitationally bound to their cosmic dance. Altogether, eighty-some galaxies form the Local Group, each one containing anywhere from thousands to billions of stars.

Another hundred million light-years pass—for the sake of argument, you have long ago stopped aging, for you are now over 100 million years old—and you finally see the Virgo Supercluster, a region of space holding more than one hundred galaxy *groups*. This massive web of galaxies contains an unfathomable number of stars—no matter which system of units we use, or what analogy we make, at some point numbers cease to make sense.

There are over 10 million superclusters in the observable universe.

The observable universe is the *only part of the universe that we can observe.* There is stuff beyond the observable universe— other stars, galaxies, and structures that we haven't yet glimpsed— but their light hasn't reached us yet, even with new, state-of-the-art telescopes like the James Webb Space Telescope. It is stuck in the cosmos, forever trapped along the creases of spacetime, the fabric of the universe, slowly making its way to us over cosmic time. Regions beyond the observable universe are still part of our universe, they are just beyond the limit of what we can observe, making them invisible (or *un*-observable).

And finally: the universe. Here it is enough to simply say that the universe is *everything, for all time.*

If we were to zoom out and see the universe—majestic, sprawling out in all directions, containing everything we have ever known and everything we ever will know—we might finally understand our location: a minuscule blip, an aberration much, much smaller than the drop of ink that went into writing this text, smaller even than the dot of an *i* on this page, smaller than the atoms forming the ink, and the protons and electrons in the atom, smaller than the up and down quarks—*the building blocks of the universe*—that create, well, everything.

All of this is to say that our cosmic address might, in fact, look something like this:

> *Number, Street*
> *City, State, Zip code, Country*
> **Planet Earth**
> **The Solar System**
> **The Milky Way**
> **The Local Group**
> **The Virgo Supercluster**
> **The Observable Universe**
> **The Universe**

You know the feeling of your stomach dropping out of your body when you're on a roller coaster? For a moment you're gliding up the track, excitement pooling in your insides as you get closer and closer to the peak, and then you're there and everything suddenly drops out from beneath you. You're in free fall, your stomach rising into your chest, past your heart, and then falling, falling, falling somewhere below.

By age ten, this is how I wake up every morning. At the time, I have no language for the sensation, other than "my stomach hurts." After the initial few complaints, my parents hastily schedule doctors' appointments, convinced the mysterious ailment is of a physical origin. We see pediatricians and specialists, but nobody can give us a clear diagnosis—for all intents and purposes, I'm a perfectly healthy ten-year-old with a strange, persistent stomachache, who just happens to get sick a lot. And it is *a lot*. By mid-February I've missed a full month of school due to illness: one week of strep throat, three different ear infections followed by a particularly brutal sinus infection, and another week of the flu.

My mom, who takes me to most of the appointments, gets steadily more and more frustrated at what she perceives as the doctors' incompetence. For every "We just don't know," she snaps at them to do better, decrying their ineptitude. After each appointment, I walk out to the waiting room while she stays behind to quiz the doctors. I learn to preemptively apologize on her behalf—her assertiveness makes the doctors (and men, in general) uncomfortable. While I wait for her, I stock up on lollipops from office waiting rooms and keep the blue wrappers, my favorites, in a drawer at my desk.

To be clear, more often than not my mom's frustrations were well placed. Doctors tended to sweep past her as though she were invisible. She would speak loudly to fight to be heard, and I'd roll my eyes in response, mortified as I watched her try to take up space.

During these appointments, she tries speaking to me in Arabic, her sentences full of hidden messages, but I pretend I don't understand. The language is a reminder to everyone that we are different, something I *hate*. At school, I'm careful to mirror the outfits that girls in my class wear, begging my mom to take me to the mall to stock up on bold-faced neon Gap sweatshirts and frayed Abercrombie jeans. From a kind-faced blond girl named Chloe, I learned what a hair straightener was, and I spend hours in the bathroom burning my hair, ironing it so flat that the roots singe. Popular girls like Elizabeth notice immediately, oohing at the ringlets turned straight. Every day, I wake up an hour early, get out my CHI straightener, and blearily work my way through my curls until my hair is passably *not* Egyptian.

My mom's Arabic belies my efforts to blend in. She speaks the language joyously, her voice lilting with laughter and music, *al-musiqa al-Arabiyah*. She speaks English beautifully, too, but says there's no joy in it. Sometimes, she locks herself in her bedroom and calls home to Cairo, spending hours laughing and yelling to her aunties and cousins. Every so often, I'll crack open the door and peek in on her calls, just to see her joy, her American mask discarded at the door.

But when we're at the doctor, that mask stays firmly on. During one exceptionally horrible appointment, I'm forced to drink a pale, chalk-white liquid that slides down my esophagus and burns my intestines. I fight nausea as I sit for the prescribed two-hour wait, the cocktail solidifying into something heavy in my guts. Finally, a nurse escorts us to the imaging room, where the MRI machine scans my body and finds nothing. We leave the office with no answers, my mom puzzled. I follow her out the door, slightly disappointed but smug, too, for I secretly kind of love being sick.

It's not that I love the *feeling* of being sick. It's the opportunity to catch up on missed sleep and rest that I love. For me, sick days are precious; my two-hour weekly TV allowance is rescinded, and

I get to watch Nickelodeon and Disney Channel to my heart's content, channels I'm typically prohibited from watching. My daily tennis practices are canceled, my diet forgotten, and best of all, my hours of homework postponed. So, I soak them in while I can, racking up more sick days than all of my friends combined, reveling in the rest I will later learn is called self-care.

One morning in September, I wake with one of my worst stomachaches to date. Eyes still closed, I reach for the half-drunk Gatorade left over from yesterday's tennis practice on my bedside table and gulp it down, willing the sugar and electrolytes to replace the nausea churning in my stomach. It tastes metallic, and I abruptly set it back down.

I toy with asking to stay home from school. For a minute I lie in bed, eyes still closed, staring into blackness as I heave and try to steady my breathing. Slowly, I blink my eyes open and, to my utter surprise, find a foreign room staring back at me.

The well-worn wooden dresser, the sparse bookshelf, a dark-red beaded macramé curtain cascading down the doorframe: none of it is familiar. I frown, and slowly realize that this isn't my room. And for a long moment, I don't know where I am.

The nausea pangs once more. Someone knocks, and a head pokes through the door. I rub my eyes and fumble for my glasses.

"Oh! You're awake! Wonderful. I made you some toast. Will you be ready to head to school in ten minutes?" Blinking away the sleep lining my eyes and straightening the frames of my glasses, I finally identify my dad's executive assistant, Annalisa, as the talking head. Rollers stick atop her graying hair, bopping with her head as she moves.

And then, I remember. My grandmother Grannell, my dad's

mother, is on the verge of death with something called pancreatic cancer, and yesterday my mom rushed to California to take care of her. My dad, still on a business trip in Laos, called Annalisa in a panic, begging for her to watch me while they're gone.

I nod mutely to Annalisa, not trusting myself to speak. She smiles kindly and shuts the door softly behind her. Dimly, I put on nondescript clothes and make my way to the car. I'm barely aware of the time or the too-bright sky as we drive to school.

Grannell is my favorite grandparent. Until recently, she lived in Dallas, Texas, just a three-hour drive away. Now she lives near San Francisco, next to my dad's sister. I miss her so much.

Before moving, she'd visit us one weekend a month, and together we'd bake snickerdoodles and put our heads together to solve puzzles. Her favorite puzzles were three-dimensional geometric ones with wooden shapes—tangrams, she called them. She handmade my favorite one, and I play with it sometimes on our coffee table. One day, while watching me stubbornly try to shove the parallelogram into place, she told me that solving puzzles is like solving math problems. I still don't understand why.

I had seen her just one month ago, when my aunt Jennifer called to say that Grannell was getting sicker. My parents and I flew out to California and drove to her new Granny Unit, and I almost got carsick on the narrow mountain switchbacks.

She'd decorated the Granny Unit in the same way as her Texas house, like a gigantic playroom. Miniature chairs the size of my palm lined a glass case in the foyer of the house. Atop my favorite white-rimmed table, a hand-carved wood toy monkey hung from a wooden ladder. When I put the monkey at the top of the ladder, it fell down, rung by rung.

Her artwork was everywhere. Hanging from the walls were several of her paintings, gigantic oil-based triangles broken into miniature geometric ones. Once, my dad told me that embedded in the pattern of miniature triangles is a mathematical solution to Grannell's favorite puzzle. He said that she used to be a programmer for JCPenney as one of their female computers, before computers were real machines.

I think she is a genius.

When I saw her, she looked like an entirely new person: skin sagging on her arms, hair white and wispy, once-sharp blue eyes now cloudy. She stooped as she walked, shuffling from one end of the apartment to the other.

During that visit, she'd asked me her usual questions: how is school going, how is tennis, am I studying hard? I'd made sure to sound cheerful, to bring the Texas sunshine to her house in the woods.

At the end of the visit, Grannell had pulled me aside. Her eyes looked sharper than they had all week.

"I need to tell you something important, Sarafina," she said to me, gripping my wrist firmly.

"Sure, Grannell," I said. "What is it?"

"My number-one piece of advice for you: make good choices. Remember that." She locked eyes with me as she said it, and time stopped for a moment as I gazed back.

"I will," I heard myself saying, my voice small. "I promise."

By fifth grade, in addition to our mandatory forty-five-minute daily chapel, we now do prayer circles in class. Usually, I pass the time by inspecting my fingernails or smuggling a book onto my lap, hidden by a sweater I throw on top, but today I raise my hand.

"Can I add to today's prayers?" I ask my teacher, Ms. Sexton, and the crow's feet around her eyes stretch wide in surprise.

"Of course, Sarafina." The eleven students in my class and I spread out across the muddy brown carpet and sit cross-legged in a circle. Ms. Sexton dims the lights and lights a candle on her desk, and I take a deep breath, inhaling the scent.

"Today I ask that we all pray for my grandmother, Grannell. She's really sick and I think she's going to die soon." I clench my friend Holly's hand in mine, trying to distract myself from my nausea. Her fingers wrap tightly around mine.

"She has cancer. I'm really, really scared." My voice breaks. I hesitate, looking up, and Ms. Sexton smiles kindly down at me, nodding for me to continue.

"I'm just . . . I'm going to miss her a lot."

Ms. Sexton perches beside me, her linen dress swooshing against my leg, and says, "God, our Heavenly Father, please grant Grannell a peaceful journey to heaven. We pray to you to relieve her of pain and regret. Amen." I close my eyes so hard my forehead hurts, careful to enunciate every syllable: *A-men.*

A round of amens follows, our tiny voices pleading in harmony. As everyone else rises, breaking the circle, I stare into the candle and once again utter *Amen,* but quietly this time, in Egyptian Masri: *Ameen.* I hold the *ee* as I watch the flame rising to touch the sky.

The chorus of prayer relieves the pain in my stomach, just a bit.

My thought when I first saw Ms. Sexton was that a witch had been brewing an age potion and something catastrophic had happened, transporting her to our yellow classroom at St. Andrew's instead. She was small but wiry and could have been forty or eighty—she was ageless in a natural way, unlike the St. Andrew's moms using plastic surgery and fad diets to attempt to rewind their biological clocks. It isn't that she wasn't beautiful: her eyes were a deep green, reflecting the plants she loved so much, and her smile an authentic thing that pierced, bit, and glowed, somehow all at once.

Her smile, now imbued with a soft kindness, shines down upon me as I return to my seat. I pull out our homework on fractions and am once again hit with nausea; this time my heart feels like it's going to crawl out of my throat. I bow my head, grimacing, and close my eyes for what seems like the tenth time that morning.

"Sarafina, may I speak with you a moment?" Ms. Sexton's voice floats through the air and slices through the undercurrents of vomit bubbling within me. Mechanically, I rise from my chair and walk over to her desk. It's cluttered with notes and pencils, old science experiments, magnets, and a mismatched bouquet of freshly picked flowers I'm suddenly sure she picked herself.

"I'm so sorry to hear about your grandmother. She sounds like a very special woman." I nod, mute. "Your parents mentioned you started therapy last week—how is that going?" I look up at her in surprise; I didn't know she knew about therapy.

"Uh." I swallow. "It was okay. She actually let me sleep for half the time, which was really nice." I blush and mumble the last bit. Ms. Sexton smiles indulgently but stays quiet, making space for me to continue.

"Honestly, we just talked about my mom for the rest of the time. And I wrote her a letter. But I'm not going to send it!" The last point feels important to clarify. My therapist says that it's okay to feel my anger and write it down, even if it feels scary. In my letter, I'd written that I didn't want to have to take care of her, that I was ten years old and still a *kid*. I'd underlined "kid" three times.

Even though I didn't show my mom, I felt better afterward. The next few days were relatively peaceful. The letter, at least momentarily, had seemed to do the trick.

"Don't worry, your secret is safe with me," Ms. Sexton says, and then continues: "I spoke with your parents and your therapist, and we thought we'd try something. How would you feel about not receiving grades for the rest of the semester?" My head snaps up, nausea forgotten.

"Wait, what?" Questions hum in my brain like radio static, their frequencies tangled together. "What do you mean?"

"I know you've been struggling with getting sick this year, and I wonder if all of your stomach issues and discomfort might be from stress." She speaks slowly, deliberately choosing each word.

"Your therapist suggested—and I agree—that we can try this to see if it helps. You'll continue completing assignments and taking exams, but you won't receive grades for any of them. The idea is that you'll get to learn without any of the stress!" I blink, trying to understand. She leans in, hair falling down the panes of her dress.

"You're an excellent student, Sarafina. But sometimes I wonder if all this"—she waves her hands toward the students bent over their textbooks, frowning—"isn't healthy. I think a break will be good for you."

"So . . . just to be clear, it doesn't matter if I get an A or an F?" I ask, narrowing my eyes.

"It doesn't matter insofar as you won't receive any grades. At all. You're free to learn, Sarafina." She smiles indulgently.

"Now, go explore."

Hardly believing my luck, I walk back to my desk and return to my papers of fractions. They don't suddenly make sense—the fraction bar is still unintelligible—but the nausea roiling inside is, for the moment, gone.

For the first time that morning, I'm able to grin.

But back at home, things just feel worse and worse. My parents' fights have grown more and more frequent over the last few years, ever since the nineteen Middle Eastern men, one of whom was Egyptian, hijacked four airplanes and crashed them into the Pentagon and Twin Towers. Immediately after the attacks, we spent the next few weeks urgently on the phone with our Egyptian family. I remember Al Jazeera News on loop in the background, and for once, my parents stopped fighting. My dad, a mix of Albanian and German, blindingly white and six-foot-two, with sharp blue eyes and sandy blond hair, looked at us with concern, but there was little he could do to protect us.

My mom had a hard time admitting that what we experienced was racism. Arab Americans, I would later learn, had fought hard to be considered white in the United States, for whiteness afforded privilege and opportunity. But it became a harder sell after 9/11, when our neighbors began calling us terrorists. One memorable afternoon, I came home from school crying because a classmate cupped a hand around her mouth and whispered with narrowed eyes, *sand nigger.* Months later, my mom was flying to D.C. for a Census conference when she was accused mid-flight of being a threat to the aircraft. They turned the flight around, grounded the plane, and kicked her off.

My parents fight every day. Their arguments have moved from the balcony to the kitchen, so explosive that they can't wait to shield me by moving the fight outside. They fight about money; who should cook dinner; when to go back to Egypt; how to raise me; the cause of my stomachache; my tennis practice; my test results; how much my dad works; how much he travels. My mom screams often, her eyes wild and pupils dilated. My dad is like an iceberg, cold and hard; callous, even. To me, he seems strong, levelheaded, and she seems dramatic, hysterical. They think I don't see, but I watch from my books and my stomach hurts.

One night, my mom flies into a rage at something I don't understand. She shrieks, her voice echoing through the halls of our house, our carpet and curtains doing nothing to mute the noise. My dad responds quietly, suggesting she go out for a smoke to calm down, and I escape upstairs with our golden retriever, Summer. I wrap my fingers around her fur as we hide in my bed, letting her lick my face to calm me down. I grab *Harry Potter and the Prisoner of Azkaban* off my bedside table, sit up, and read it to her, pretending we're siblings who are waiting out the storm together. Her hot dog breath is comforting, a reminder that she's still here and she loves me.

Some hours later, my mom lets herself into my room. My bed creaks a little from the collective weight of the three of us: me, Summer, and her. I close my eyes to pretend I'm asleep and feel her fingers lightly caress my face. I concentrate on keeping my breathing even, intent on fooling her, when I feel a drop of hot liquid, then another, and all of a sudden, my eyes fly open and she's looking down at me, hair wild and tears falling.

I get up then, letting Summer stay asleep at my feet as I slide back against my headboard.

"Are you okay, Mom?" Normally I would let her do the talking, but she looks so defeated that I waver.

"No," she chokes out before continuing, "I don't know what to do. I'm so miserable. I hate it here; I hate this. This marriage is broken." Her eyes are wide and her pupils dilated.

Against the black of my room, she looks little more than a ghost. Her coiled curls are in disarray, water dripping from the ends as though she'd just gotten out of the shower. The molten passion that fuels her is gone; what's left is a broken shell, cracked and fading.

I feel so small in this moment. The bed stretches out beneath me, wide and far, until I feel as though I am being swallowed. I try reaching for her hand, but it lies limp within mine.

"Maybe you two can try couples therapy?" I suggest, desperate. But I'm not desperate to help; I'm desperate to make the conflict stop. I want my mom to hold me, to check in and make sure I'm okay, to love on me.

Her gaze is soft.

"That's a good idea. Maybe . . . maybe I'll try that." Her voice wavers. But this is a way out and she sees it too. That light at the end of the tunnel affords her a semblance of hope, and she grips on to it just as I do, two fools hanging by a thread.

"I think it's a good idea to try. It will help you two get to a better place."

She hiccups and wipes the crusted salt from her face.

"Maybe you're right." She curls up next to me, and I slide my feet to touch Summer, aching to feel her fur instead of my mom's skin. I grip it and let the bristles poke in between my toes as my mom drifts asleep. My stomach rumbles, not its usual ache but with something like resentment. I lie there, wide-awake, for what feels like hours.

Ms. Sexton's favorite supplementary activity to our classically Episcopalian education was to take us on nature walks. She'd organize us in a messy line, unlike my first-grade teacher's strict orderliness, and place herself at the forefront, clucking like a mother duck at her chicks.

On the afternoon I talk with her about grades, we take one of her famous walks. Ms. Sexton says nothing as the line relaxes into a jumbled mess of fifth graders, allowing herself to be wrapped into the center. She whistles tunelessly, and I trail after her.

Holly, one of my friends in class, elbows her way through the cluster over to me.

"How'd it go with Ms. Sexton earlier?" Concern furrows her brow. Her blond hair shows its highlights in the sunlight, and I briefly lose myself in their glimmer before responding.

"It was okay. Don't tell anyone, but she said she's not going to give me grades anymore because I keep getting sick! Stress, or something." I roll my eyes to show I didn't ask for this solution, and at her expression I let myself grin.

"Wait, what? That's amazing!" Holly high-fives me (discreetly), and we wait for the students around us to pass before continuing.

"Does this have to do with your therapist?" Her eyes widen behind her glasses, the farsighted lenses magnifying them into bright-green jackfruit. She's the only person I confided in about the appointment. I nod.

"Yeah. I mean, I don't think I really did much in therapy—" I laugh. "But I guess something is coming out of it."

She nods wisely. "I bet that's it." And then she laughs, not unkindly. "You're such a worrywart."

Sighing, I mutter, "I know." She squeezes my arm, smiles again, and then rejoins Anna and Nadia. I spot Ms. Sexton wandering along the side of the path and speed up to catch up to her.

I count the leaves on the passing trees, noting which ones have red berries I can pick later. My skin, turned cinnamon-brown by Egyptian melanin and hours of tennis practice, hungrily soaks up the sunlight. The rays are warm as fire and soft as velvet.

Walking, I discover, is enjoyable when you have nowhere to be. My breath comes easier when I'm outside, the oxygen infinite.

Ms. Sexton and I walk aimlessly, the rest of the class fading behind me.

"Do you see that tree over there?" Ms. Sexton asks, finger stretched out. I look to where she's pointing and spot a massive cedar. Its branches are buckling under the weight of hundreds of ripening berries, their scent infusing the air with a woodsy, gin-like aroma. She inhales deeply.

"People hate these trees. They say they're water hogs and noxious weeds. They want to completely remove them from the Hill Country. Eradicate them." We arrive at the tree, and she inspects the branches thoughtfully.

Dropping a branch, she looks back at me, her gaze deadpan. "It's all old wives' tales."

"Oh," I say carefully, unsure how to react. She looks at the tree intimately, me and the class seemingly forgotten.

"People like to create stories about things that suit them. They blame other people for their problems; hell, they even blame *trees*!" Her head swivels back to me, and my heart shudders for a moment under the weight of her gaze. She takes a deep breath and sighs.

"Cedars are resilient. People mistake their strength for danger. Their berries can be used for medicine and tea. The branches themselves are homes for native wildlife. The trees are drought-resistant, taking little from the land"—she pauses for a moment— "or at least, as much as other plants in the region. They get a bad rap.

"When you look at a tree like this, what do you see?" She tilts her head curiously.

"Uh." Unsure what the right answer is, I nervously knock my heels together and shift my weight from hip to hip.

"Just observe, Sarafina. What do you *see*?" By now, the class has gathered around the tree, forming the day's second circle.

The wind whips into our shirts and caresses the leaves dangling in front of us.

"Lots of leaves and branches," Anna, third best friend to me and Holly, says.

"Good. What else?"

"Berries and sap. It gets all sticky on the branches," Elizabeth, a girl I admire for her popularity, says. The heads in her vicinity nod in agreement.

"Okay, sure. What else?" Ms. Sexton is looking at us intently, as though this is the most important lesson of the day.

"I notice that there's lots of them. But also, that each one is different. Some are tilted, others knotted. This one is really straight and tall," I say, speaking quickly at first and then slow by the end.

I find myself thinking, for no particular reason, of the sky. How it is everywhere all at once, how it is blindingly infinite, how it makes me feel like I'm not alone. And then I think of the stars, and how they are like the leaves on this tree, many in their distinct glow but one in the cosmos.

Ms. Sexton nods, satisfied. "Yes. We are all different, just like this tree. We're all growing, like the branches stretching out with berries and sap. And we're all part of something beautiful. There are many of us, and we are all unique." She points at Elizabeth, and then me, and then the one Black boy in the class, Andrew. "That uniqueness makes us—just like this tree—beautiful."

A couple of boys, Will and Patrick, roll their eyes, and I hear their elbows jostling on the other side of the tree. But I stay quiet, transfixed. Holly and Anna stay quiet too.

"This"—she gestures a little wildly—"is science. We are observing the beauty of the world around us to learn a little more about our planet. To learn a little more about each other, and ourselves." That monstrous anxiety inside me, licking fire along my guts just this morning then quieted by Ms. Sexton's gift, is musing. My nausea is replaced by curiosity, and it feels like there's enough space for all the stars in the sky to fit within my chest. The new-found cavity is liberating, novel.

The tree towers in front of us, imperious. When we file back to the classroom, I turn around and spot a hummingbird perched on its branches. Its wings pause while it rests, drinking the sap from a knobbed cavity in the wood. It gorges itself on the sweet-ness of the tree and then flies off, wings beating in unison as it returns once more to its home in the sky.

Despite the generous reprieve I get from my anxiety monster, this is the afternoon I have my first full-blown panic attack.

Annalisa picks me up from school and, after I fold myself into the backseat, informs me that I need to call my dad.

I've just taken out the math pages from my three-pronged binder, intent on making use of the forty-five-minute drive to practice.

"Here, you can use my phone," Annalisa says kindly. I take it out of her hands and grip the heavy thing while she tells me which numbers to press.

"Hello?" My dad picks up on the first ring.

"Hey, Dad, Annalisa told me to call you. Aren't you in Laos or something?" For a long, pregnant pause I hear nothing, and then—

"Honey, I'm so sorry to have to tell you this. Grannell just passed away." The words hang, cracked and pulling apart at the seams, my brain scrambling to make sense of them. His voice is alien to me. Annalisa looks into the rearview mirror sympathetically; she must have already gotten the news. It is moments later when I realize I'm supposed to say something.

"I'm sorry." My statement hangs in the air like a question. Death is unfamiliar. I'm unsure how to compose my words, my thoughts, my flesh. I sit paralyzed in the seat, my sweat staining the leather, soaking the seat with anxiety and fear.

I'm faintly aware of my dad talking through the phone, and only catch the end of his sentence. He must have been speaking awhile. ". . . flying there now. It's going to be okay, little one. I promise."

I nod, then realize he can't see my head moving.

"Okay. Yeah."

"I love you," he adds, his voice so full of love that I break a bit.

"I love you too." The line goes dead, and I am entirely lost.

The math on the page swims before my eyes. My chest tightens and my stomach aches. My monster is back. Shuddering, I gulp at the air like a fish.

"Are you okay?" Annalisa peers back through the review mirror at me.

Where is all the oxygen?

Trees, I look at the trees outside the window. But they're moving so fast, the car speeding past them, that I can't decipher any of their leaves. Their branches paint a line across my window, vaguely menacing the faster we go.

"Um." I can't get words out. I raise my hand to my face and am surprised to feel hot tears staining my cheeks, leaving trails of salt in their wake. I dig fingernails into my left palm, hoping the acute bite will overshadow the crippling pain in my chest.

To my dismay, it does nothing.

I know, *I know* I'm dying.

My thoughts cycle in one terrible loop: *She's gone, forever. It's permanent, she's never coming back. I should have been kinder to her, a better granddaughter. I should have been supportive to my dad on the phone. I am a bad daughter. I'm going to be alone. My dad is going to die one day. I'm going to die one day.*

Hot acid burns my stomach, and I inhale a faint scent of that terrible chalk drink.

The pain is so immense that I *want* it to consume me. I want to feel all of it—I *deserve* to feel all of it—and yet it is unbearable; I want none of it.

Annalisa's voice, quiet at first and then insistent, cuts through my vision. "Sarafina? Are you okay?" She meets my gaze and sees something unfathomable there, for she immediately stops the car on the side of the road. I press my hand to my chest to massage the knot squeezing my heart.

She gets out and joins me in the backseat. She slips her body next to mine, letting me collapse onto her. Her knuckles are white gripping mine, but the pressure helps. I am sobbing now, racked with tears as I shake in her embrace. She is saying words, but I no longer hear them. We sit there, I, small in her embrace, she, towering and gracious, together in that one-person seat, for what might be minutes or hours.

The worst thing about panic attacks is that you feel like you are dying. You struggle for oxygen, your chest cavity heaving with the weight of ten thousand worlds. Adrenaline floods your bloodstream and spikes your heart rate and you gasp, convinced your heart is failing. Your skin turns to fire and then ice, goosebumps erupting like tiny little mountains, forcing you to shake so hard you fall to the ground. Your heart stutters and thuds unevenly in your ears.

Of course you're not actually dying.

You're living.

But isn't living the hard part?

CHAPTER III

FORMATION

Humans are fascinated by stories. We tell stories for levity, for play, to gain insight and comfort. Most of all, we tell them to learn a little more about ourselves. Stories have a beginning, a middle, and an end. Most times, a story is a neat little thing that can be tied up with a ribbon and bow. We nod at the end of one, when the hero defeats the monster or the princess saves the prince (heteronormative gender expectations aside, I like this version better), and we say, "Yes, that's how it's supposed to end."

The formation of our solar system and its planets follows a similar arc.

The beginning: Some time ago, about 4.5 billion years, an enormous cloud floating in the universe accumulated so much gas and dust that gravity pulled on it hard, forcing it to collapse. While we're not sure *exactly* why the cloud decided to collapse at that exact moment—was it due to a passing shock wave triggered by a nearby exploding star? Did it collide with another interstellar cloud? Or was it the result of some random gravitational pull?—the cloud collapsed nonetheless, spinning up as it contracted into a spiraling disk.

The now-spinning cloud disk got *hot* as it smushed all of its contents into a more compact region. Temperatures were hottest and pressures highest at the center of the collapsed cloud, where the most material gets compressed. It is there that tiny elements got so hot that they started careening into one another, slamming together, emitting light, and then forming heavier elements. This process is called *nuclear fusion,* and it's the process by which most elements are formed in our universe. By igniting nuclear fusion, a baby protostar later became the center of our solar system: our sun.

In astronomy, interesting features are often "clumpy." It's a silly word, but it gets the job done, and it's exactly what happened inside the spinning cloud. Small dust particles began gravitating toward one another and clumping together, forming larger and larger clumps until they were large enough to be called *planetesimals* and then even larger *protoplanets*. These are baby planets, the precursors to the planets scattered across our solar system today.

At this point, everything cools. The disk fragments and the protoplanets and protosun contract until they have cooled enough to finally be called planets and our sun. Our solar system as we know it today has been born.

As you might recall, the inner planets of our solar system are rocky, *terrestrial* planets. They have solid surfaces covered with rock and dust that we can land spacecraft on and explore. On the other hand, the solar system's four outer planets are gaseous, without solid surfaces. Which raises the question: why are some planets rocky and others gaseous if they all formed from the same cloud at the same time?

The answer comes back to the center of our solar system, the sun. Terrestrial planets are so close to the sun that the heat quite literally blows off any semblance of a substantial gaseous atmosphere. Sure, the inner planets have some atmosphere—on Earth, we would die without one!—but nothing compares to the giant planets that harbor enormous hydrogen and helium atmospheres. These atmospheres dwarf their inner cores, enveloping them in a massive blanket of gas.

Are there other solar systems out there like ours? Are there other *planets* like ours? Or are we unique, destined to float through the cosmos like a lone blue speck cast into a bottomless ocean?

As we search through the cosmos for an answer, we're increasingly finding other solar systems not dissimilar to ours. They might be bigger or smaller, have two suns or three, but in principle they're more or less the same. Some have *hot Jupiters*, or gaseous planets that likely formed at the outskirts of their system and then migrated inward, toward their sun(s). Others have dozens of *exoplanets*—planets orbiting stars outside of our solar system. These differences are beautiful. They add texture and richness to what a solar system can look like, and what it can be.

But we have yet to find another Earth. We've observed planets close enough to another system's host star where it's neither too hot nor too cold to host liquid water, a necessity for life as we know it. Planets in this *habitable zone* are deemed *Goldilocks planets*, those that happen to orbit in this special, unique zone, where it's just right.

And we keep looking. Instead of being a random, unique fluctuation in the fabric of the universe, it seems increasingly likely that there are other solar systems and other planets similar to ours. We cast our telescopes up toward the sky, and our spacecraft out into the abyss, for a chance to better understand just how alone we are.

The ending to this story is still unfolding. We are perched at a precious moment in time, a climax to the plot of our story, while we explore the very fabric of our universe. As we learn more about our origins, we learn just how unique the Earth is, how special it is that we are here.

We learn about ourselves.

Science camp probably doesn't sound all that fun for most people, but when Ms. Sexton tells our fifth-grade class that she leads the two-week summer camp, I immediately beg my parents to let me go. After they agree, I spend the final few months of the school year fantasizing about camp, daydreaming of days exploring the Texas Hill Country and nights under the stars. I imagine wading through creeks and looking up at the Milky Way, unencumbered by the stress of my parents' fighting or the anxiety of school. The prospect of exploring for exploration's sake is intoxicating.

Despite my excitement, the first of June creeps up on me. The days have started to get hotter, and Texas is already claiming strings of weeks that climb beyond one hundred degrees. That morning, my mom and I drag my oversize trunk to the car and pile in, sweating against the suede cushions as we drive toward the middle of nowhere. My dad is gone on yet another business trip, so my mom is tasked with the responsibility of dropping me off at camp. Together, we sit in the car awkwardly, and she turns on classical radio to fill the silence.

I glimpse wide expanses of Texas ranches through the car window as we leave civilization. Foremen with brown, wide-brimmed hats and ornate leather cowboy boots cajole bulls into doing their bidding as their horses gallop around and pause to munch on barrels of hay. I glimpse a family of goats trotting around an arena, and a particularly obstinate one pawing at the entrance, refusing to enter. We pass by lakes of blue and hills of green, and I am entranced, staring unblinkingly at the scenes outside. The visions of another world—lush, slow-moving, alien to me as an urban Texan—flash by as we drive on.

Three hours later, the car slows, and we pull up a rocky, unmarked road in front of an enormous log ranch house. Wooden rocking chairs are scattered across the front porch, slowly swaying to the song of the wind. A bird feeder dangles from one of the log pillars, and a hummingbird is fluttering nearby, drinking a golden liquid from one of the trees. It is idyllic, so much so that I am taken aback. I can feel my baseline layer of anxiety slipping out of my body and evaporating into thin air as I step out of the car, and a sweetness taking its place. Even though I've just arrived, I immediately feel a sense of release.

I grab my trunk and wheel it up the long side ramp as my mom waits outside the car. The inside of the house is austere, mostly empty, with a wide common room. A long table—for dining, I assume—stretches the length of the floor, and behind it is a plain kitchenette. A poster hangs from one of the walls, with CHORE CHART in capital bold Sharpie letters scrawled across the top. I recognize Ms. Sexton's handwriting, the curly *T* giving her away. I am oddly comforted by this, knowing that she's here.

Ms. Sexton appears in front of the kitchenette just as I look around, searching for a map or some indicator of where I should drop off my trunk.

"Welcome, Sarafina!" The outdoors agrees with her; her normally white skin appears salted, like leather that has been treated and tanned.

"So happy to be here." I grin. "My mom's waiting outside, so I just need to drop off my trunk and quickly say goodbye." She waves her hand again, this time kindly dismissing my rush.

"Take your time. We're all just getting settled in. And give her a hug—she'll miss you!" I roll my eyes at this, more to myself than anything, but nod. She points me in the direction of the girls' room and walks back to the kitchenette, humming to herself.

The girls' room is brimming with bunk beds. Three girls have already claimed the pair of bunks along the back wall and are perched at the edges of their trunks, unpacking. A small, slight one confidently introduces herself as Brittany, and I think to myself that she's going to be the leader of the group. Her brown hair is straight, her skin beautifully white, and her teeth perfect. I look at her with a pang of jealousy, immediately aware of how not straight my curls are, how brown my skin is, how crooked my teeth are. But I smile with teeth back at her, hoping to hide my insecurity.

Her tall, gangly friend Suzanna empties her pillow and toiletries kit behind her. She tucks her ginger locks behind her ears as she tilts her head in greeting. She's quiet, letting Brittany do most of the talking, and allows stray red strands to fall in front of her face.

The last girl is Priyanka, my best friend from home and next-door neighbor. I run to hug her, and admire how we look together, two slightly different shades of brown; I, slightly taller, and she, slightly more petite. We coordinated our matching outfits, and I smile at the mirror image of myself: boys' athletic shorts from Target; oversize T-shirts, mine from tennis and hers from soccer; and baseball caps to hide our hair. Any nerves I felt at meeting the other two girls immediately disappear. Priyanka hugs me back, and we negotiate the bunk-bed situation: I call the top bunk

(*It's technically closer to the stars! Plus, I'm older.*) and she resigns herself to the bottom.

After pointlessly shuffling stuff around in my trunk, I finally stop procrastinating and look at Priyanka.

"I'll be back in a minute, I still need to say goodbye to my mom," I mutter. We exchange a knowing look, and I stuff my spare sweatshirt into my trunk before disappearing back outside.

Mom's still standing by the car and looking at the trees scattered around us. Blades of switchgrass are so tall they come up to my thighs and tickle me when they sway in the wind.

My mother is watching the wind kiss the grass, mesmerized. She rarely looks like this: outside of herself, calm and at peace. Her eyes have a faraway look, and I'm not sure she registers where she is, or even who she is. Interrupting feels intrusive, so I stay by the porch for longer than I need to. She is unaware of my presence and continues looking at the plain stretching out in front of her.

I can feel the time passing as sweat pools up underneath my brow, and finally decide to join her by the car. She looks at me with a faint trace of reproach, followed by anguish. She quickly disguises both with a smile.

"Well, honey, I'm going to miss you. Two weeks is a long time!" Her eyes well up with tears, and the eyeliner, kohl, rimming the bottoms of her eyes is already getting smudged. She rubs her eyes, smearing the kohl some more.

My stomach flutters, and I'm suddenly eager to leave her behind and rejoin Priyanka, Brittany, and Suzanna.

"Yeah. I'll miss you, too, but it's just two weeks. It's not that big of a deal," I say, rolling my eyes. I want to muster up empathy, but it feels inaccessible, hidden somewhere beneath my anxiety.

My response is so obviously deficient that I'm not in the least surprised to hear her sharp inhale, and I brace myself for her reproach. To my surprise, though, in the lull, she masters her anger and looks at me once again, this time in sadness. But her vulnerability makes me more resentful, angrier. I think back to that letter I wrote in therapy but never gave to her: *I don't want to have to take care of you.* I feel justified in my anger, and I give it permission to fill me up.

"Right." She gives a little laugh. "That's a long time in Mom years." Her eyes crinkle, tears welling up once more. I'm quiet, unsure what to say.

"What am I going to do without you?" she asks, and at this I glare. What is she to do? Is she not an adult?

"You'll be just fine, Mom." My voice is hard. "I gotta go, camp is starting, like, now." I reach out my arms to give her a hug, eager to hurry along the process, which she reluctantly accepts. She stands there like a cardboard cutout, completely still under the weight of my arms. My resentment builds, but it's covering up something hidden more deeply.

I want her to hug me back. I desperately want her to hold me and tell me that she wants me to have fun, to explore, that she can't wait to hear my adventures when I return. Her encouragement feels like a double-edged sword—for every encouraging push, she acts like I'm abandoning her, leaving her alone to muddle through the pain of her isolation. I don't yet understand the intersection of immigration and loneliness, nor the friction between brownness and white suburbia. But I feel her tension and I resent her for it; I want her to be strong. Able. Supportive. Independent.

And at the same time, I want her to blend in.

I don't yet realize that I want those things in her because I want them in myself.

Science camp is nature's playground. We spend the first few days of camp exploring local streams, identifying animals, and classifying them in little blue spiral notebooks. We pretend we're pioneers, the first people to ever spot the hidden pathway to the lake or the soft patch of grass under the aged, knobby cedar tree at the entrance to the camp.

I'm awoken every morning at five o'clock by birds dancing outside our room's windows, chirping songs to welcome the day. The sun is already rising, and the moon is beginning to dim. I spend those precious moments alone in bed, staring up at a ceiling that I can almost touch. The relative quiet during that time, aside from the twittering singsong outside my window, is like nothing I've experienced before. Even though there are four of us sleeping in beds stacked on top of one another, the room feels spacious in those early mornings. The hot coals in my stomach don't have time to simmer; they stay cool, placated in a restful peace.

By eight o'clock we make our way to the common room, where we join the four boys sleeping in the other bunkroom. Whoever is on breakfast duty has already laid out the cereal and yogurt along the long dining table at the center of the room, and the other girls and I always rush to get there as soon as possible to beat the boys to the limited provisions.

After breakfast, Ms. Sexton stands at the head of the table and walks us through our plans for the day. Some mornings we get to meet animals on the ranch and others we do trail maintenance. Every day after lunch, we meet in the field of grass behind the ranch house to do "meditative nature journaling" in our notebooks. Ms. Sexton gives us a prompt and instructs us to sit in silence for the next thirty minutes, reflecting on what we saw that morning and how it made us feel. The rest of the campers take advantage of this time to secretly send notes to one another, but I quickly learn that these half-hour increments are some of my favorite times of the day. I choose a spot slightly away from the rest of the campers and begin to write, sometimes about a particularly beautiful flower I found or a spider I saw, but sometimes about things I hadn't yet seen that day, like the night sky, my anxiety monster, or my parents.

During the afternoons, Ms. Sexton leads biodiversity classification lessons. We trek through the forest with her, learning how to navigate by compass and identify animals. She teaches us the differences between damselflies and dragonflies, both of which are readily abundant at this time of the year, and walks us through classification techniques. I learn that the animal I'm most afraid of is the brown recluse, a particularly venomous spider with a dark, violin-shaped mark on its back. I investigate each rock before I sit down, expertly looking through the nooks and crannies to make sure no spiders are hiding, waiting to strike.

Priyanka, Brittany, Suzanna, and I have formed a club of girls unafraid to do anything. After leading us through species classifications, Ms. Sexton allows us free time to explore together. We spend late afternoons in the lake, swinging from the rope swing and backflipping off the dock. Brittany teaches me how to do a back dive, and one afternoon I practice so many dives that I'm nauseated by the end. We rest on the sun-stained wood, drinking in the sunlight and drifting off to the gentle slap of the water into the dock.

The Saturday after our first week, about halfway through camp, the four of us decide to explore a forgotten stream in the far east side of the woods. We pack a backpack with trail mix and chewy chocolate-chip granola bars, deet mosquito spray, and sunscreen, and agree to take turns carrying it. Our blue spiral notebooks are stacked together at the top for easy access. Suzanna grabs a bandanna to shield herself from the sun—she burns easily, we found out—and I grab a disposable camera. The four of us wave goodbye to the boys and Ms. Sexton and make our way outside.

It's a hot day, already ninety-five degrees even though it's hardly ten o'clock. I immediately start sweating when we get outside and am suddenly grateful that my mom insisted we pack more tank tops than T-shirts. I'd begged to stick with boys' clothes—my standard uniform included oversize T-shirts and long basketball shorts—but she insisted I bring the tank tops too. I liked how boys' clothes fit for comfort and ease of mobility, while girls' clothes were made up of frilly lace and shades of pink. Before camp, I'd told my mom this was stupid, and she'd laughed in a rare moment of solidarity. It was annoying, she agreed, but tank tops serve a purpose too. I guess she was right.

I lead the way as we follow a natural path carved by rainfall and animal hunting tracks. After five minutes, Suzanna shushes us.

"Do you see that?" her voice rings in alarm. Her arm is outstretched, index finger pointing to a cranny in the woods I have to squint at to make out. With all the trees and bramble, the spot is almost entirely hidden.

"Oh my God!" Brittany shouts, and then claps a hand over her mouth. Suzanna shoots her an angry look as Priyanka and I move forward, straining to see. I still can't see what they're talking about, until some branches move and Priyanka mutters a quiet "Oh!"

All of a sudden, a coyote appears, staring at us from about fifty yards away. Both of its ears are perked, one flopping downward and the other entirely straight. Even from fifty yards, I can see that its eyes are trained directly upon us. Brittany utters a small yelp, and Suzanna grabs her hand. Priyanka and I stand there, staring at the animal as it stares back.

I take half a step forward and Priyanka looks at me, eyebrows raised. I take another half step and she follows suit.

"What are you *doing?*" Brittany whispers, panicked. I stretch out my palm behind me to indicate for her to stay, and I continue walking forward.

I taste salt, presumably from the sweat dripping down my cheeks. Leaning into instinct, Priyanka and I crouch down as we walk to look unthreatening. But after another three feet, the coyote cocks her head once more and immediately bolts.

Brittany and Suzanna catch up to where we're crouching, and the four of us break out into a fit of startled laughter.

"You are insane, Sarafina. Totally insane," Suzanna blurts out in between laughs. I'm laughing, too, charged with adrenaline.

"Okay, but did you see her?! How beautiful was she!" The coyote with her small, lithe body, her eyes wide with curiosity, was stunning. She was studying us, just as we were studying her. Her deciphering gaze was so fierce that it was uncanny, almost human.

Priyanka nods, absentmindedly kicking a branch in front of her, then swivels around to look each of us in the face.

"I hope she's okay. I mean, it must be really uncomfortable to have humans invade your home."

"What do you mean?" Brittany asks.

"Well—" Priyanka gestures at the creek behind us, the spiders crawling up an overturned rock and a small rabbit peeking at us in between blades of grass. "We're in *their* space. We're guests here. It must be scary to see us and not know if we're going to hurt them."

Suzanna frowns, and Brittany fiddles with a stick. Her shoulders slump with guilt.

"You're right. Now I feel kind of bad . . . I wonder if getting closer made it scared?" I feel a little embarrassed that I didn't realize this when she points it out.

"It probably didn't help," Priyanka agrees kindly. The monster of anxiety within me yawns and stretches, deciding whether to wake up.

There was something so intoxicating about exploring the unknown, but she was right, and I knew it. I remembered one of the lessons Ms. Sexton told us on day one of camp: *Do No Harm.* I interpreted it as respecting my fellow campers, and we'd all looked at one another with hidden grins when she listed this rule—of course we were going to be nice to one another. But I realize now that I'd misunderstood.

She hadn't just meant that we should be kind to one another; she meant that we should respect and honor the space in which we were exploring. It is a privilege to be in someone else's home. No matter how big or how small, how violent or how kind. This is their domain, and we are simply guests, lucky enough to see the world through their eyes for a brief moment in time.

"Okay, how do we fix this?" Suzanna asks pragmatically.

"I'm not sure we can. But let's just make sure we leave the trail as we found it," Priyanka suggests. We all nod.

"Yeah, of course. Want to sit at the creek over there and rest a bit?" Brittany points to a stream of water behind us, clear and beautiful, gently rolling across the landscape. We follow her over to a cleared patch of soil and sit down in the sun, silently admiring the eddying water streaming past us.

Damselflies flit around, skimming across the top of the water. They're rainbow in color, azure blues and frank oranges, gorgeously reflecting sunlight as they hop from crest to crest. I'm mesmerized by how liberated they are in movement, how daringly they zoom through the air. I wonder what it must be like to feel that free.

"You know, the color of this water reminds me of the Red Sea in Sharm El-Sheikh," I muse. "We went there for my birthday last year, and I got to snorkel in the sea. It was so pretty, but I think these damselflies might be even prettier colors than the fish I saw." I say this without thinking, more to myself than anything, but immediately pause at the end. Priyanka knows all about my family in Egypt, my annual trips to Cairo and my birthday spent by the Red Sea. But I haven't yet shared this with Brittany and Suzanna.

Since third grade, when the Twin Towers fell and an Egyptian was found to be one of the terrorists, I'd kept my homeland to myself. We'd learned to stop delivering Christmastime baklava, our version of cookies, to our neighbors when one slammed his door in our faces, causing the pastries to slide off the painstakingly decorated plate and onto the ground. Will, one of my classmates, still calls me a terrorist when he thinks our teachers aren't watching.

To my surprise, Brittany and Suzanna smile widely.

"That sounds so cool! How beautiful," Suzanna exclaims. She's looking at me in awe.

"Seriously, it's so neat you went to Egypt," Brittany says admiringly.

I'm taken aback. "It is?" I ask, raising my eyebrows.

"Dude, extremely cool," Brittany says reverently. "Egypt is so exotic." She draws out the *o* in "so," infusing it with her preteen excitement.

I frown a little. "*Exotic.* I guess so."

The word "exotic" feels weird on my tongue. I massage it a little, sounding it out as though it's the first time I've said it. It's not one I've considered to describe Egypt. "Exotic" sounds like an adjective for the coyote we found, or for aliens on another planet, not a word to describe my aunts and uncles, family dinners and spice markets—*home.*

But despite my discomfort, I notice that they're still eagerly looking at me to share more. To them, the idea of Egypt is sexy, strange, and enticing in its unfamiliarity.

"It's my favorite place in the world. I'm actually half-Egyptian, so we go over there a lot to see our family. My mom was born and raised in Cairo." I mumble the last part, still a little afraid of the judgment I expect will come.

"That's so cool! Does your family ride camels to school?" Brittany blurts out. I break out into laughter. For all my unease, I'm relieved to be able to share this part of myself with them. It's a strange feeling, this intimacy.

"No, they drive like we do. But I did get to ride camels to see the Bedouins in the desert."

"Who are they?" Suzanna asks, opening up the backpack to chew into a granola bar. She looks totally at ease, reclining against a massive cedar and munching on the bar. Brittany follows suit, sliding next to her and resting her head on Suzanna's legs.

"They're called Awlad 'Ali—nomads who live in the desert. We visited them one day and got to learn more about how they live. They had a lot of sheep." I laugh. I unconsciously touch my hand to the bony part of my neck, where a Bedouin woman gave me an Eye of Horus to wear for protection. I'd left it at home, along with my golden scarab and my necklace with the tiny pyramid, afraid of getting made fun of. My heart pangs: I wish I'd brought them.

Priyanka throws a leaf at me, interrupting my reverie.

"Sorry, what?" I ask sheepishly. She nods her head toward Brittany, who flips her hair and then continues.

"I *said,* that sounds really cool and I want to go! My family is so boring, I've never actually left the country." Brittany rolls her eyes.

She digs her toes into the soil and kicks some dirt up at the rest of us. Priyanka giggles and kicks some back her way, and Suzanna jumps up, grabs a fistful of water, and sprinkles it on top of Brittany's head. All of a sudden, the four of us are scrambling into the creek, splashing water at one another and laughing uncontrollably.

I dunk my head underwater and close my eyes, feeling around for the stones, smooth beneath my feet. When I resurface, I feel light, buoyant. It's not just the water but my insides too. Looking around at the three girls, I am amazed by how at home I feel, under this canopy of leaves, wading through a stream trickling through the middle of nowhere.

In this brief moment my insecurity about simply being me disappears. There's something about the wilderness that can strip our differences away, until all that's left is the core of us. There's a rawness in the exhilaration I felt in seeing the coyote, and a deep sense of self I recognized at Priyanka's observation. The ease in sharing my home, Egypt, with these girls—Brittany and Suzanna as newfound friends and Priyanka as she'd always been—was shockingly comforting.

As the days go on, I have a hard time remembering to feel out of place in my skin or about my hair when we're trekking through creeks and rock beds, pulling algae from our toes and mosquitos from our skin. And when I think about that coyote and how majestically she stared at us, a look of alien confidence and human understanding, I am calm. How wonderful it must be, to move so confidently through the world. How beautifully freeing.

The time passes in a peaceful haze, and when Ms. Sexton reminds us that we've made it to the last day of camp, I'm shocked.

"But it's Wednesday!" I shout in dismay. Priyanka turns toward me, eyes wide. Ms. Sexton frowns for a moment and looks at me, perplexed.

"What? Honey, it's Saturday. You've been here for thirteen days." The campers laugh and I frown, baffled.

"Thirteen?" I'm stunned, and at my visible confusion Ms. Sexton winks.

"Time flies when you're having fun, huh?" One of the boys, Chase, whoops and the remaining three girls laugh. Brittany shoves him flirtatiously, and he grabs her hand.

It feels like someone dropped an anvil in my stomach, my dread so pronounced I can taste it. The prospect of returning home is excruciating.

"I have something that might help though. Sarafina, are you listening?" Ms. Sexton asks pointedly. I have a blank stare on my face, thinking instead of my parents and the inevitable math problems awaiting me at home for "summer practice." Priyanka stamps on my foot to get my attention, and I wince.

Ms. Sexton continues, unperturbed. "A special guest is going to visit tonight. It's a real treat that he's coming."

Chase raises his hand, and instead of waiting to be called upon shouts, "Michael Jordan!"

"What? Chase, this is a camp for science, not basketball." The boys laugh and I roll my eyes.

"No, an astronomer from a big university is going to visit. Do y'all know what an astronomer is?" Ms. Sexton asks. The group stays silent, but I'm suddenly paying attention.

"An astronomer? Like, a person who studies the stars?" I blurt out. For a moment I'm sure I imagined her question, for her face is unreadable.

"Yes, quite. His job is to study the night sky to learn more about the stars, planets, and other celestial phenomena. He's going to bring his telescope and show us what some of these things look like."

A smile blooms across my face as her words sink in. An astronomer? Here? The years of listening to *StarDate* echo in the back of my mind, reducing the chatter around me to a faint, gnat-like buzz.

"—so take the rest of the day to secure your belongings and say goodbye to whichever special places you found during the last two weeks. We'll reconvene after dinner under the awning outside to meet our guest and do some stargazing." At this, Ms. Sexton turns to me, eyes glimmering. As she's walking away from the circle of campers, she pulls me aside.

"Get there early, Sarafina, and you can ask him some questions before the rest of the group arrives," she suggests shrewdly, winking. At that, she turns back around and walks out of the room.

The four of us girls return to the bedroom to pack up our trunks. Each of us secretly slides notes of friendship into each of the trunks before snapping them shut. The sadness of our impending departure is palpable, and Suzanna spontaneously hugs Priyanka halfway through gathering her things. Once we're done, we look around at one another, unsure of how to spend the remaining part of our last day.

"Let's go down to the dock," Brittany suggests.

A pang of sadness bites into me as we follow her down to the water. I don't want to leave. We move slowly, savoring the hike and the scent of the woods we'd gotten to know so well. I take long, deep breaths, inhaling all of the calm I imagine floats in between the branches of the trees, in hopes of bringing it back home with me.

When we finally get to the lake, we pause.

"Oh God, we forgot our swimsuits," Suzanna groans. Priyanka shrugs, and without missing a beat runs and leaps into the lake fully clothed. We stare at her for a second, then start laughing and follow her lead, T-shirts and all.

The four of us swim around, dragging our feet through the sand and lying on our backs soaking in the sunlight. Priyanka strikes up a game of Categories, but I'm too distracted to pay full attention, and by the time I've lost three games in a row they give up on me. I swim back to the dock and pull myself onto the faded wood, trying to center myself.

I'm fighting excitement and dread as the night nears, torn between looking forward to the astronomer and anticipating how awful tomorrow will be. The minutes slide by, so slowly that I squint to look at the sun over and over again, waiting for the moment when it disappears behind the tree line. The other girls have long ago stopped splashing and trekked back up the path to shower and prepare for dinner. I stare into the sky wistfully, wishing I could float on this dock, with waves gently lapping its sides, forever.

Finally, I look up and the see only trees silhouetted in twilight. The colors of the sky reflect off the water, pale reds cresting and breaking against the shoreline. I do one last dive, savoring the way the cool water lightly caresses my skin, and head back to the ranch house. The gentle haze of twilight casts a light fog on the trail, and damselflies dart in and out of the leaves as I walk on.

After a quick shower, I don my last clean pair of shorts and oversize tennis-tournament T-shirt and glide through the dining room to snag a sandwich before making my way to the porch. My heart is already thumping, and goosebumps erupt along my shoulders despite the balmy eighty-degree night.

Once outside, I spot the silhouette of a man bending over a large object in the middle of the field. He's adjusting something and pulls hard until a lid falls off. I squint, and belatedly realize this is the telescope.

The stars are bright against the black of night. Silvery shards of light reflect against the metal of the telescope, making it appear as though it were conjured by magic. I've never seen one so big; it's at least as tall as I am and as wide as my torso. My jaw falls open and I start jogging over to him, raw excitement fueling my steps.

"Hi!" I yell. He whips around, startled, and I slow down.

"Sorry, sorry, my bad. Didn't mean to scare you. Hi! I'm Sarafina, a camper here." I speak quickly, slightly out of breath from my sprint and excitement. It's too dark to really make out his face, but I think I detect a frown.

"Okay. Hello," he says. I reassess, pausing to give him space. I must have frightened him. I try a different tactic.

"Um, are you the astronomer? Ms. Sexton suggested I come out and meet you before the rest of the campers join. She knows I love astronomy." I offer a small smile, allowing my face to plainly show my enthusiasm.

My eyes have started to adjust by this point, and I begin to distinguish parts of his face—the solid, straight line of his mouth, his down-turned eyes, the bitter wrinkles in his brow. I take a slight step back.

"Yes, I am," he says flatly. His curt responses sting, but I'm looking at that telescope and I *want it.* I try again.

"I want to be an astronomer, too, when I grow up!" I'm already envisioning it: Older Sarafina, perhaps twenty or thirty years old, bent over a telescope and witnessing the sky and stars up close. She's confidently solving math problems in her little blue spiral notebook, drawing perfect diagrams and expertly directing the telescope to her next target. Books pile up around her, and she flips through their pages, knowingly circling phrases and incorporating them into her next measurements. She's fierce, like the coyote. Confident, self-assured, and smart.

He looks me dead in the eyes, the red light that he's carrying in his right hand lighting his face.

"No, you don't. This is not for you." He says this silkily, his thin lips curling in derision. Despite his facial expression, I freeze, sure I misheard.

"Sorry?" I ask quietly.

"This. Is. Not. For. You," he repeats, reveling in the pauses between each word, and continues. "Astronomy is incredibly difficult. Some of the smartest people I know struggle with it. Hell, it's even hard for me."

His gaze sweeps over me, and he sneers.

"But I love it," I offer in a small voice. He snorts in response, turning back toward the telescope.

"So what?"

I stare blankly at his jerky movements, watching him prepare the telescope for viewing as though I'm no longer there. He peeks through the eyepiece several times, muttering to himself and rotating knobs. The longer I stare at him, the farther away he appears, until all of a sudden, I blink and I can barely make him out. Water drops on my hand, and I realize I'm crying.

The sound of the rest of the campers floats through the air, and I begin to make out Priyanka's laughter and Chase's low voice. Suddenly sure that this is the last place I want to be, I twirl around and slide past the group of them, hoping to disappear back into the house without anyone noticing.

To my chagrin, Priyanka grabs my arm as I move by, and before I say anything she reads the expression on my face.

"You're not okay," she whispers matter-of-factly. I can't get words out, so I just shake my head, tears still falling. She grabs for my hand and together we walk side-by-side back toward the house. Ms. Sexton calls for us, and Priyanka yells over her shoulder that I'm not feeling well. She supports most of my weight as we walk, until we get back to our room and I collapse onto her bed.

We sit in silence for a while as she lets me cry, and I curl into her on the bed. Finally, when the tears have slowed, I recount what happened.

"What the hell! That should never be allowed!" she says furiously. I just nod along miserably, gripping the sheets to keep my hands from shaking.

"We should tell Ms. Sexton. He's awful! This is awful!" She's yelling now, and I shake my head frantically.

"No! Please, no. I don't want her to get mad at me. He's her friend, I think. I don't want to ruin that." I am pleading at this point, grabbing Priyanka's wrists. She looks at me for a long while, unblinking.

"Fine," she concedes, and I exhale a long sigh of relief.

"But he's not right. You can become an astronomer if you want to," she declares authoritatively. I heave another sigh and look at her.

"How do you know he's not right?" I whisper. I rack my brain trying to understand why he doesn't like me—is it because I was too excited? Too young? Not smart enough?

"Because I know you. And because you love it," she says. I rub my eyes and give her an exhausted smile.

"I hope you're right," I whisper.

"I am," she insists. We exchange a small smile and lie next to each other in bed, staring at the ceiling and pretending like we're looking up at a sky of stars.

The next day, we get up early to make Ms. Sexton a surprise good-bye breakfast. I'm still shattered by the astronomer's comments, but Priyanka's words now echo alongside his in my mind. When Ms. Sexton finally walks in and spots her breakfast set at the head of the table, she smiles broadly, a soaring expression so full of happiness that my disappointment is briefly forgotten. Despite the terrible interaction, I'm able to muster a smile. Science camp was still nothing short of a dream.

"Thank you all for a wonderful two weeks. This has been one of the best camp experiences I've ever led. I hope you all learned a thing or two." We nod reverently, and she continues.

"I want each of you to know that you'll always have a place here. And if you ever miss this place during the year, remember that you have the outdoors right outside." She winks and tucks into her cereal. The campers start talking amongst themselves and laughing through their goodbyes, Ms. Sexton's speech already forgotten. I sit without an appetite, stirring my cereal, until it's time to clean up and I drop my bowl into the sink.

Ms. Sexton glides up behind me and rests her hand on the counter.

"Are you feeling better?" she asks. I blink, wondering what she means, then remember Priyanka's excuse for ushering me away so quickly last night.

"Yeah," I say, pasting on a grin. "Just needed to sleep. I'm going to miss this place so much!" I look around the now-familiar kitchen, through the window just above the sink, and my smile grows, no longer feigned. It's so beautiful here.

Parents are already lining up outside to pick us up, but I dilly-dally in the bunkroom to prolong my departure until I hear Ms.

Sexton yelling for us to hurry up. I slowly drag my trunk outside, heaving it across the wood floors.

The sun blinds me when I get to the patio, and when I blink, I'm surprised to see both my mom and dad standing together, waiting for me. My dad is holding my mom's waist, and they're smiling.

"Honey! Hi!" My mom bursts into tears, and my dad, face flushing with embarrassment, turns quickly toward the car, tugging the trunk along behind him.

"Hi, Mom, hi, Dad. Missed y'all." My mom hugs me so tightly that I gasp for breath, and then she looks down at me, smiling with tears in her eyes.

"I missed *you*! Tell us everything." We pile into the car and I wave goodbye to Ms. Sexton from the backseat. I'm still waving as the ranch house fades out of sight.

"It was really fun. I love being outside. And I got to meet an astronomer." I add the last part quickly, but both parents turn around grinning.

"That's amazing, little one! How was that?" my dad asks, still smiling. My heart aches. I pause, deciding whether to share the astronomer's painful words, but when I glance again at the front and see my parents humming together, smiling as they wait for my response, I realize I don't have a choice.

"It was good. The stars are beautiful out here." At the very least, the second part of the sentence was true. My parents continue smiling and asking questions, probing for as many details as they can get.

As I stare at them, watching as they enjoy this brief moment of peace between us, I find myself thinking once more about the astronomer's words. But this time, instead of believing him, I remember Priyanka's protestations, Ms. Sexton's encouraging smile, those days spent exploring the wilderness.

And the thoughts careen into me: *He was wrong. I can do it. I will.*

For the rest of the drive, the three of us swap stories from the last two weeks, and I nestle into the backseat, inhaling the remaining tendrils of forest scent.

I can do it, and I will.

CHAPTER IV

EVOLUTION

Since the dawn of humanity, people have been drawn to the unknown like moths to the light. To reach the depths of the oceans we learned to breathe underwater; to travel the cosmos we learned to fly. We have spelunked in subterraneous caverns and summited the tallest mountains on Earth, sent robots to distant planets, and walked on the moon. It is in our DNA to be curious. For much of humanity's existence, we have explored as we evolved; we sought out water sources to satisfy our thirst and created a means to make fire to warm us in the cold, fashioned tools out of bone to forge our way through the world, and constructed houses from ice and wood to protect us from the elements. But beneath the utility of these discoveries is a singular human desire to know oneself—and, by extension, the universe in which we live. In braving the unknown we seek answers, allowing our curiosity to light our way through the cosmos.

Over the past seventy-five years, advances in technology have propelled humans into the far reaches of the cosmos, pushing the boundaries of our knowledge of the universe. In fact, spacecraft are already strewn across the solar system. At the time of this writing, our own moon is home to four lunar orbiters from the United States, China, and India. A lander from China roams the moon's ten thousand craters—and that's just the number of craters on the near side of the moon—crawling up and down the pale-ivory rock to study everything from the solar wind to lunar topography. These spacecraft gaze upon the Earth from their vantage point in the night sky and bear witness to Earth's phases (just like the monthly lunar phases, but from the opposite perspective) as the moon orbits our home planet. Our collective bravery carried us to the moon, allowing us to walk its surface and look back upon the Earth, our fragile home planet, which from the perspective of the moon hangs like a marble in the vast blackness of night.

Only two spacecraft have ever flown by Mercury, with a third on the way. The *BepiColombo* spacecraft is currently en route to Mercury, the closest planet to the sun, to deliver two orbiters that will float around the planet to study its internal and external composition and structure. When they arrive, they'll find a scorching-hot surface coated with a wispy atmosphere of light elements like hydrogen and calcium. Enormous plains dotted with massive volcanoes cover the outer surface, churning up elements deep inside the planet's exceptionally massive iron core. *Mercuryquakes* (think Earthquakes, but on Mercury) crack the surface, forming fissures and vents that spew lava across rolling plains and hundreds of impact craters. It is an inhospitable and infernal planet, the shortcoming of forming so close to the massive ball of gas and fire that is our sun.

Venus is home to one active and several failed spacecraft missions and is a hellish planet of (mostly) dead robots. It is similar to Earth in some ways: it has roughly the same mass, size, and density, but that's where the overlap ends. Venus is Earth gone wrong, its damned twin sister. On Venus, the stench of rotten eggs saturates clouds of sulfuric acid, creating cloud formations so thick that they entirely obscure the planet's surface. Mists of poisonous sulfur rain down upon a yellow, smooth surface covered with active volcanoes. Venus is an unforgiving place. The crushing atmospheric pressure, over ninety times that of the Earth due to Venus's extraordinarily dense atmosphere, obliterates any real chance of a human surviving on the planet. Venus is odd: It rotates the wrong way, a possible result of being struck early on in its formation by a large clump of material. Or it could simply be that the heavy atmosphere is so bloated that the sun's gravity tugs hard on it, flipping the planet from a conventional rotation. And as if the searing heat, overpowering pressure, and poisonous gases weren't inhospitable enough, any oceans or rain likely evaporated when the opaque atmosphere blocked thermal radiation from escaping, a process called a runaway greenhouse effect. The planet's temperature now averages around 900 degrees Fahrenheit. Venus offers a forewarning of what can go wrong if we do not protect the Earth.

Mars, the planet aside from Earth that we have most explored, is home to a whopping eighteen orbiters and fifteen surface rovers. The total combined weight of artificial objects on Mars is approximately 9,470 kilograms, or about the mass of a large elephant. If humans were to live elsewhere in the solar system, Mars is at the top of the list. It is unusual in its ruddy color, a result of iron oxide (rust) that blankets a beautifully rocky surface of volcanoes towering far above sweeping canyons as big as the continental United States. Martian dust coats massive sand dunes and is carried across the planet by swirling winds and planetwide dust storms that obscure a pale-pink sky. Make no mistake: Mars is a hellhole (as journalist Shannon Stirone so eloquently put it). But despite the unforgiving temperatures, which can plunge as low as −200 Fahrenheit, punishing winds, and tenuous atmosphere, Mars enchants us, for it contains a vital ingredient for life: water.

At some point in the long-distant past, perhaps hundreds of millions of years ago, salty liquid water flowed across the Martian landscape in serpentine channels. The meager atmospheric pressure, less than 1 percent that of Earth, makes liquid water impossible now, but Mars's frozen subsoil and icy frozen terraces hold enough water to cover the entire planet once over. Long ago, Mars might have been hospitable, when it had a thicker atmosphere and oceans of liquid water. But now, Mars is also a cautionary tale for Earth—if it used to be livable, what went wrong? Or was it doomed from the start because of its small size? Should we fear the same fate for Earth? And if so, how can we prevent a planetwide catastrophe?

While we have yet to land a spacecraft on Jupiter—in large part because it's impossible; the planet has no solid surface— nine spacecraft have flown by it. They found a strange world: one with at least sixty-four moons, a sort of miniature planetary system; giant-size lightning storms of "superbolts" that light up a surface of volcanoes; turbulent storms raining down true diamonds onto clouds of gas; an underground ocean of mushy, metallic hydrogen—all on a planet so large that one thousand Earths could easily fit within its volume. Jupiter has been cooling since its formation, causing the planet to shrink and jacking up temperatures within the core. Although Jupiter is nearly 500 million miles from the sun, the planet's interior gets scorching hot, reaching temperatures as high as 60,000 degrees Fahrenheit.

The giant world of gas is surrounded by strange, exotic moons. Io is home to thousands of volcanoes, hundreds of which are erupting simultaneously, spewing directly out into space. Similar to Venus, it stinks of rotten eggs. The surface is coated in a burnt-orange layer of sulfur compounds, which cover a subsurface ocean of liquid sulfur and poles of frozen sulfur dioxide. Volcanic plumes are thick and opaque, obscuring the many impact craters that dot the satellite world. It is a harsh, violent moon of rotting gas; a world embedded in transformation as volcanoes spew, destroy, and then cool, reshaping the moon's surface every few months.

Europa, another Jovian satellite, is like a shattered marble. Dark stripes that look like necrotic veins spread out across the surface, perhaps formed by ice fractures cracking the surface into a million different shards. Below the icy crust lies a subsurface ocean, one that reminds us of Earth's many liquid saltwater oceans. Giant plumes of water vapor float through the crisscrossing stripes of ice and escape into space. Europa is an intriguing moon—thanks to its liquid saltwater oceans, it is currently our best chance for extraterrestrial life in the solar system. We look at Europa longingly, with the hopes and dreams of beings searching for extraterrestrial life in the vast expanse of the cosmos. The Europa Clipper, launching in 2024, is poised to answer this very question.

Over 400 million miles beyond Jupiter floats Saturn, a planet of exquisite rings and even more extraordinary moons. Four space probes have been sent to Saturn, famously capturing Saturn's stunning ring system: a collection of ice formations, some as small as a grain of dust, and others as large as a London double-decker bus. They extend over 75,000 miles and orbit at 45,000 miles per hour, encircling the planet with rings within rings alongside eighty-two Saturnian moons. Geodesic patterns ripple across the planet as clouds form in the shape of a hexagon, tearing across the planet's north pole. Saturn is similar to Jupiter, with its gaseous nonsurface and mushy metallic hydrogen interior coating a core of rock and iron-nickel. Violent winds whip across the pale-yellow planet at impossible speeds, reflective of massive storms brewing across its surface.

However, what is most fascinating about the Saturnian system may not be the planet itself but rather two of its moons, Titan and Enceladus. Titan, the largest moon of Saturn, is just under half the diameter of Earth and has the soft-yellow hue of a dandelion due to its thick methane-and-nitrogen atmosphere. As the only other known body in space (aside from Earth) to have liquid on the surface, Titan offers an extraordinary look into the possibilities for life. Winding rivers and pools of methane and ethane lakes loop around tactile dunes, and methane ice volcanoes jut out of the surface. Just as the water cycle loops through various phases on Earth, Titan's cycle does so with methane.

About one-tenth of the size of Titan, Enceladus circles Saturn like a spherical mirror of ice. Geysers jet water vapor into space, which then snows back down onto the icy surface, coating smooth plains and cryovolcanoes spewing water vapor and organic molecules. Below the surface lies yet another ocean, this time with a possible energy source that could conceivably harbor conditions for life. Future robotic spacecraft missions are already planned for the Saturnian moon, one of our most promising locations for some sort of extraterrestrial life.

Beyond the many moons of Saturn lie Uranus and Neptune. They are desolate planets, with gaseous surfaces and surface pressures so extreme they squeeze atoms into diamonds. On both worlds, wild, frenetic winds storm across the blue-green surfaces, the result of atmospheres composed of a toxic methane gas and hydrogen soup. They are not just gas giants but ice giants too—volatile, damning worlds of wind and storms.

Two spacecraft, *Voyager I* and *Voyager II*, have already left our solar system behind and are currently speeding into interstellar space at a breathtaking 30,000 miles per hour. They offer perhaps the most profound reflections on our place in the universe, for they are traveling headfirst into the beyond. Although we can still communicate with them, there is no return journey for these two spacecraft; they are doomed (or blessed?) to float forevermore into the abyss. Now *Voyager I* is the farthest human-made object from Earth—ever.

The spacecraft I've mentioned are but a fraction of the missions sent to other worlds. In their travels, they represent the best of us: the teamwork that goes into mission development and execution; the ingenuity of humankind; the collective ideas we dream of achieving. Beyond their extraordinary science capabilities, they remind us of the human explorers who work so hard to push the envelope. It is not simply the engineers and the scientists but the artists too; the writers and teachers, politicians and advocates. It is the dreamers.

In Austin, Texas, if you have the financial means to send your child to a private school, you send them either to St. Andrew's or St. Stephen's.

Where St. Andrew's is rigid, taut as Jesus on the cross, St. Stephen's is pliant, flexible. Juxtaposed with St. Andrew's, even the campus is laid out in stark opposition. St. Stephen's is a campus on a hill, built around the chapel at its center. While this implies a certain devoutness—and it's true, St. Stephen's by name is inherently religious, an Episcopal school too—it has a different relationship with religion from St. Andrew's. Chapel services are diverse in structure and leadership. During the month of Ramadan, Muslim community members lead the thirty-minute services, describing the beauty of their religion and its practices. Jewish community members lead during Hanukkah and Passover, and Hindus join during Diwali. On Lunar New Year, the Chinese club puts on an hourlong celebration during chapel, full of brightly adorned dancers and wildly decorated yo-yos, multicolored spectroscopic streamers and traditional animal costumes.

When my parents suggested that I leave St. Andrew's and transfer to St. Stephen's, they were surprised by how readily I agreed. When I visited to tour the campus, I was captivated by its sheer size, how students freely walked around in groups of twos and threes, some climbing trees at the edges of manicured lawns and others sitting in circles on the grass, laughing and comparing words in their notebooks. The sense of freedom, one that St. Andrew's seemed to suppress, was palpable as I walked the campus; I spotted teachers taking their classes outside to enjoy the sunshine, and one group walking down to the forest behind campus, nicknamed the "Gulch," to identify leaves growing just beyond the campus boundaries.

Best of all, St. Stephen's had an astronomical observatory. It was one of the first things I noticed when we drove the long, winding road onto campus, rotund and stark white, mushrooming between the cedar trees encircling the main campus. It was huge, wider than a classroom and taller than St. Andrew's chapel.

It was perfect.

My parents' justification for moving schools was twofold: For one, St. Stephen's had a nationally renowned tennis academy, and they were intent on me seriously pursuing the sport. The second reason was more obscure, and one I only realized after I'd graduated and moved on: they were concerned (though they'd never tell me outright) that St. Andrew's put a stopper to intellectual curiosity, fostering anxiety instead.

Regardless of their reasoning, I was thrilled.

My first day of sixth grade, I walk into the main room of St. Stephen's Middle School. It's in the center of the building, with couches randomly scattered across the floor. A hallway to the left is lined with classrooms, and the dean and administrative offices are off to the right. On my campus tour, Mrs. Douglass, the dean, called the main room a "common room" and I smiled to myself. It reminded me of Harry Potter and the common room in Gryffindor Tower where Harry, Ron, and Hermione would meet to wade through their homework and relax by the fire. The name gave this room a magical quality, and I imagined a crackling fireplace at the opposite end of it, instead of the floor-to-ceiling panes of glass overlooking the Gulch.

Mrs. Douglass is standing at the center of the room handing out class schedules to new students. I look around and spot two boys, clearly already best friends, wrestling over their papers and tackling each other. A teacher to the side of the room clucks her tongue and sharply tells them to quit it. They smirk and walk away, jostling each other as their heels clap against the concrete floors.

I walk up to Mrs. Douglass and hesitantly extend my hand to receive my schedule. Her steel-gray hair is pinned up atop her head, and a few loose strands escape to frame her face. She has kind eyes, and I say my name a little more confidently upon spotting the laugh marks that wing her eyes.

"Nance, Sarafina," I say. She frowns slightly and peers down at her list, scanning the names with her index finger.

"Sarafina . . . El-Badry? Nance? Is that right?" she asks, stumbling over "El-Badry." I grimace. My mom must have included it when registering me for school, and my heart pangs angrily. She tells me to be proud of my middle name—her maiden name— but I hate it. It's alien, foreign, and people mock it, laughing as they observe what a weird sound it makes. They ask me why there's a hyphen, and what it means, and who has a name like that anyway?

I blush. "Er, yes. You can just forget the middle name and just say Sarafina Nance," I hurriedly try to correct her.

"Why? You have such a beautiful name!" she says, smiling. I stare at her, mute, sure I misheard before remembering to respond.

"Um, thanks. It's hard for people to say, so I don't use it." I fidget with my watch, itching to grab my schedule from her fingers and slip away.

"I think it's unique, just like you. That's beautiful." She smiles kindly once more and pointedly looks away as my eyes unexpectedly well up. I sniffle and wipe my sleeve across my face as she flips through her pages.

"Ah, here we go. Sarafina Nance." She hands me the paper, and my eyes are mercifully clear. I scan it quickly before turning away and pausing.

"Yes?" she asks, her eyebrow arched in waiting.

"This might be weird, but St. Stephen's has an astrophysics class, right?" I ask.

At this she frowns. "Yes, in the Upper School."

Despite the doubt sweeping across her face, I push. "Is it possible to add it to my schedule?" My hope buoys me, and as I imagine swiveling the telescope to look up at the stars, little goosebumps erupt across my arms.

"You're a sixth grader, Sarafina," she says, huffing a laugh.

"Right, I know," I say quietly. "I just thought I'd try."

"I admire the passion," she says, still struggling to swallow her laughter, "but you'll need to wait. You can take it as an eleventh grader. Until then, stick with the classes we've assigned you. They'll be challenging enough. This curriculum is designed to push you in ways you've never been pushed before. You're here to learn how to think and communicate, analyze, and develop arguments." She winks, and then loudly calls for the next student to come grab their schedule. I'm shoved to the side by a pointy elbow and find myself sitting on one of the couches.

I stare blindly into the crowd, my hands gripping my schedule hard so that I don't cry again. The anxiety flooding my body remembers the astronomer at camp, the one who so easily sought to crush my dreams with a sentence.

But as I sit, I notice another voice inside me, one that is quietly taking up space.

She's the voice of the stars. She is infinite and small, somehow both at the same time. She is calm. She doesn't speak, but she pushes me, slightly against my heart, to remember the sky. To remember that it's out there, waiting. It's not going anywhere. And neither am I.

Throughout that first day of school, it's drilled into us that academic expectations are high.

"You won't succeed if you're passively learning. You're here to be an active thinker and to learn to ultimately be an active member of society."

Mrs. Duren, the math teacher, takes a slightly different approach by telling us the sheer number of hours we should expect to spend on homework. "If you're not spending five to six hours on home-work each night, you're doing something wrong."

A few people roll their eyes, but any levity in the room is squashed when she hands out a pop quiz. It's a page full of math I've never seen before, problems made of letters instead of num-bers. My stomach clenches as I work my way through each one, biting back the tears threatening to fall.

By lunchtime, I'm exhausted. By four o'clock pickup, I'm so overwhelmed by the newness of it all that I'm blinking back tears. When my dad pulls to the front of the car line, my head tingles with relief, and I drop like a weight into the backseat, finally able to leave the school day behind.

Before the new school year, my parents negotiated switching drop-off and pickup duties behind closed doors, so that my dad now gets to shuttle me across town to and from school and tennis practices, and my mom has more time for work. When they told me the new arrangement, I failed to bite back a smile. Afternoons and nights with Dad was more than I dared hope for, the most time I've gotten to spend with him since I could remember.

When my mom spotted my grin, she strode out of the room and slammed the door, muttering under her breath that I was abandoning her. I froze, confused what I did wrong, and kernels of doubt sprang up like weeds.

Still, despite her anger and my self-doubt, nothing could tem-per the excitement cresting through me in unexpected, giddy waves.

My parents' relationship was falling apart anyway. On our last vacation, they'd fought so much that by the end, I'd decided I hate vacations.

In the airport security line on the way home, my mom screamed at my dad that she wanted a divorce. She threw her boarding pass at him, in a line of dozens of people, grabbed her suitcase, and stalked away. I was left standing next to my dad, who tightly gripped my small hand in his.

When we finally got home, my dad left for a full day to "give her space." With him gone, she raged alone in her room, and I could hear shoes being thrown from all the way down the hall in my own bedroom. The worst part was when I heard her call my dad, screaming that their marriage was broken and dysfunctional. When they fight, anger and fear, doubt and resentment swirl around within me so rapidly that my stomach feels sick.

I wished they could either get along or get a divorce and end this torture.

One hour later, I found her downstairs in the kitchen, scrubbing Tupperware and humming tunelessly to herself. I stepped quietly, hoping to avoid alerting her to my presence, but Summer jumped up to lick me, and my mom swiveled around.

"Oh, hi honey. Don't spoil your appetite before dinner." She smiled sweetly and turned back to the sink. I brushed past her to grab the water, nodding mutely.

Summer followed me back to my room and we spent the rest of the afternoon curled up together in my bed, my hands in her fur and her nose in my chest. I cried desperately, filled with roiling tides of grief and fear and confusion. By the end, when my tears finally dried on Summer's fur, she sat up, nosing my hand for food, and I felt entirely empty. I numbly walked back downstairs, having to remind myself how to step with one foot in front of the other, and stared at my mom as she cleaned the table. A cool anger descended my spine and I embraced it, letting it shield my hurt as the two of us sat down to eat. My dad's chair seemed to stare at me from across the table, empty.

I was no stranger to my mom's yo-yoing spurts of wild rage.

But I'd been noticing them becoming more and more common. Whenever she'd rage, I'd freeze. I'd imagine an impenetrable ice-berg, painting the hard edges and sleek white contours crystal clear in my mind, admiring how it remained solid for ages, refus-ing to melt. The image would calm me, steadying my emotions as her storm swirled around me. Sometimes, when I'd be caught off guard and she unexpectedly launched into a tirade, I'd pinch my wrist while she spoke, harder and harder until she flamed out. I'd focus on the tingling pain, and it would help me survive. Still, by the end of each of her episodes, I'd learn to hate myself just a little bit more.

In the car ride home on that first day of St. Stephen's, I tell my dad that I failed the pop quiz in math. Shame, metallic and bitter, coats my words, but my dad interrupts.

"Okay, how's this for a plan? Before school every day, we'll come early to review your homework and any questions you might have."

I pause. "Really?" Even more time with my dad—I can't be-lieve my luck.

"Of course, honey. We'll get through this together, okay?"

I laugh through a sob and let the tension melt out of my body as I lean limply back into the seat.

Knowing I'll have my dad helping me along the way settles my nerves far more than a perfect grade ever would. And knowing that he's here with me settles my deeper fears, as well—the ones about my mom, and their relationship, and that maybe it's al-ready fractured beyond repair and we're falling away from each other, faster and faster every moment, until we're so far apart that we wonder if we were ever together at all.

Over the next several months we establish a familiar routine. My dad wakes me up at six A.M. and cooks breakfast: two eggs, a piece of toast, one handful of berries. I push it aside sleepily (*I'm not hungry!*) and we fight a familiar argument until I finally capitulate and shove the food into my mouth. We get into the car, and I sleep or study until we get to campus.

Once there, he parks in an empty parking lot—nobody's there at seven—and we walk together up the hill to the breakfast hall. Boarding students mill around, grabbing orange juice and eggs and blearily shoveling food into their mouths while my dad and I grab an empty table. I pull out my homework for the day, and he checks it, quizzing me as he goes.

My mom seems to (at least momentarily) resign herself to my busy schedule. Most days, I don't see her until I get home from school and am rushing around the house, stumbling to find my tennis shoes and running out the door to get to practice on time. Afterward, we eat dinner together, but once the last bite is gone, she disappears back upstairs to her bedroom. Silently, I lug my backpack to the kitchen table, where I begin a long night of studying.

Weekends are now full of tennis. Both parents join in on tournament weekends, and sometimes there are rare moments of shared joy between the three of us, especially when I win. We learn that I am moderately talented, and I make it to the semifinal and final rounds of several tournaments in a row. On a good weekend, my victories carry us all the way back home.

During the school weeks, a new friend, Emilia, often joins us for the morning review sessions. She and I build up a competitive rapport, and I learn that I love to win, to be right. She's brilliant and hardworking and these early morning sessions quickly become a highlight of my day.

While I quickly learn that English and geography are my strongest subjects, math and science remain my hardest. It's not enough to know the steps through a problem; you have to be clever. And when I'm anxious, my mind gets stuck, and my thoughts swirl, and nothing about me is clever.

Just before Christmas break, after a particularly grueling breakfast during which I declared my hatred for math, my dad joins me in the common room after a morning review session and flags down the dean, Mrs. Douglass.

From behind, I watch as they file into her office and the door slams shut. He hadn't told me he needed to speak with her, and my stomach lurches uncomfortably. I grab for a sheet of math homework I haven't reviewed, desperate to fill my mind with a tangible distraction.

"—just not as confident as I'd like her to be in math." My dad's low bass carries through the door, all the way to where I'm perched. Silently, I slide the paper back into my folder and straighten upright, rigid.

"I hear she's doing relatively decently in math, Mr. Nance," Mrs. Douglass says soothingly. "I'm not too worried, but please just remember that it's not wholly unusual that she's struggling."

"What do you mean?" I blink, stunned by the razor edge of his question. In my mind, I am picturing him sitting with one foot crisscrossed over the other leg, the other heel tapping the floor impatiently, causing both legs to wobble with a nervous energy.

"Well, you know," says Mrs. Douglass, like she's pointing out the obvious, "girls aren't really cut out for that sort of thing." She says it matter-of-factly, and in response my stomach swoops and then falls.

I stare unblinkingly ahead, like I can see the tension I imagine is building in her office, an opaque cloud of noxious gas that seeps through the door cracks and floats into my nose.

"Right, okay. Well, thanks for the chat," my dad mutters. I hear chairs scraping the ground and papers sliding around on her long desk, and suddenly the door is thrown open. My dad is standing right behind it, face drawn up in hard-edged lines of anger, and he turns to where I'm sitting.

Abruptly, he asks, "Ready?"

I jump up, nodding, and he sweeps me under his arm. We walk in lockstep through the common room, through the double doors, all the way back outside. In front of a rock outcropping lining the landscaped garden skirting the building, we pause, my dad muttering louder and louder.

"Absolutely ridiculous thing for her to say. Who does she think she is?" he says out loud, fuming. The normal calm, steady force staying his temperament is gone; I rarely see him like this.

"Dad, are you okay?" My stomach swoops again, this time in response to his anger. Did I do something wrong?

He peers down at me through his glasses, blue eyes shining brightly. "How much of that did you hear?"

I debate for a moment to lie, and then think better of it. "All of it, I think." I'm a bad liar, and an even worse one with my dad, and he knows it. He nods and inhales deeply.

"I like Mrs. Douglass," he begins, choosing his words carefully. "But even smart people can sometimes be wrong." He looks up at the trees and blinks, removing his glasses and rubbing his eyes before continuing. "And she's wrong about this." He replaces his glasses softly on his brow and refocuses his eyes on me.

"You need to know that you can do whatever you set your mind to, Sarafina. Whether it's math, tennis, becoming an actress, or a dancer." I roll my eyes and start interrupting to let him know that I don't want to do either of the latter two, but he holds up his hand to stop me.

"Listen. The only thing—the *only* thing—that can hold you back is yourself. Do you hear me?" I look up at him, eyes wide at the weight of his words, and nod. He forces a smile, reaches for a hug, and half laughs, letting the tension escape his body.

"Sometimes people can't separate their own insecurities from reality. I bet Mrs. Douglass doesn't like math very much." He winks, squeezing me into another hug, and then pushes me back through the door. As I walk to my first class of the day, my mind swirls, and that anxiety monster in my insides is actively champing at the bit.

My dad's reassurances help, but Mrs. Douglass's words stick with me, like a burr caught on the fleshy bits of my brain. And I realize, the closer I get to my classroom, that what my dad said is actually *keeping* the burr there, the glue cementing it to my thoughts.

I might be able to do whatever I can dream of, but if I don't succeed, the only person to blame—to hate—is myself. And that knowledge sinks in my stomach like a heavy anchor, dragging me down along with it.

The last day before winter break, I skip tennis practice to hang out with friends after school. One of them, a girl named Lillian, tells me she has a crush on Sanjoy, one of the two Indian boys in our math class who have joined our friendship group, and we spend the afternoon laughing together, coming up with more and more ridiculous ideas to get Sanjoy's attention and make him fall in love with her. We strip the common-room sofas of their cushions and line them up along the floor, perfect shock absorbers for jumping up and down. The prospect of two weeks away from my grueling school schedule is filling me with such relief that I'm buoyant, allowing my joy to carry me from one laugh to the next.

For the first time in my life, I don't feel like I have to pretend to blend into a friendship group at school. Lillian, American-born Chinese; and Emilia, Mexican American; and I share our unique cultural backgrounds, and over the last several months we've bonded over how comforting it is to finally be open about them with friends. We get into classic preteen dramatic fights about boys and friendship, but still, this type of group dynamic is wholly new—I never once feel judged or weird for my brownness.

At school, it's the first time in my life I feel accepted for me.

I carry our laughter and joy home that evening. In our yellow kitchen, I'm soaking in the sunshine of the room, smiling faintly as I sit down at the table, and, for once, ignore homework. But to my surprise, my dad sits down with his usual pencil, paper, and calculator.

"Dad," I say with a giggle. "It's Christmas break, remember?"

Summer walks up to me and I scratch her fur, grinning at my dad's mistake and the intoxicating possibility of doing whatever I want over the next two weeks. My only daily commitment is six hours of tennis practice (except for Christmas Day). Still, this new-found freedom makes me want to laugh and cry at the same time.

I'm looking down into Summer's eyes, a warm liquid chocolate color that melts my insides, when my dad responds.

"I spoke with Mrs. Duren and she told me that she gave everyone an extra-credit math packet. Is that right?" he asks pleasantly.

I shrug, my fingertips still deep in Summer's neck fur. "I guess, but it's not due until the end of the year, when we graduate." Still oblivious, I move onto the floor to scratch Summer's belly. She rolls over and pants, tongue lolling as I find her favorite spot.

"In that case, now is a perfect time to get ahead and master some of the concepts you've been struggling with, don't you think?" I look up to see the sharp blue of his eyes trained on me. Adrenaline floods my body, and I freeze.

"You don't understand, it's, like, fifty pages of problems, Dad. We haven't even learned some of the material yet! I think she just gave it to us now so she didn't forget. No one is starting until later."

"Fifty pages of problems is an excellent opportunity to practice and improve. Since you don't have any other homework right now, we can take advantage of this break." I can see my almost-freedom slipping away from me, evaporating into the air of this sunshiny room and out into the winter air. I feel my eyes well up with tears and I screw up my face angrily, refusing to let my tears get the best of me.

"I don't need to do this! This is my first break in forever!" I'm shouting, and my mom walks into the room to see us like this: me, hopping up from the ground; Summer, detecting the change in atmosphere and escaping into the living room; my dad frozen, except for twiddling the pencil in his right hand.

"What's going on?" my mom interrupts. I feel my stomach drop at her entrance, and my heart stutters, speeding up.

"He wants me to do extra credit over break," I say bitterly. My dad's face is unchanged, and his lack of emotion makes me angrier. "Like, fifty pages of it."

"Peter, there's no reason why she can't take a break. She's exhausted. You're pushing her too hard." My dad's face looks like it's cast from stone, immovable as she speaks, but relief burrows into me. Despite our complicated relationship, I know that my mom will protect me when she can.

I nod along as she speaks, until she says the next part and it slides into my chest like a dagger.

"She doesn't need to be a workaholic like you." Her tone has a poisonous edge to it, and I watch my dad recoil. Summer peeks at me from behind the couch, and I wave to her miserably. She stays back, scenting the tension sinking into the room.

"She doesn't have to do it all in one day; we can break it into chunks over the next few weeks. I just think this is a good opportunity for her to get ahead and show Mrs. Duren how committed she is." He says it slowly, in a level voice, to diffuse the mounting hostility.

"How committed she is? Just like you, working nonstop without regard for anything else in life? Do you realize how toxic that is?" They say all of this openly, and I want to curl up in a ball—I deeply regret bringing this up at all.

"Why don't we talk about this outside, Samia?" my dad says, and half stands up before my mom yells.

"I'm fine!" She's shouting now, and her voice rings in my ears. "I don't need a break—I'm not the problem here!" My leg hair stands up, and familiar goosebumps erupt across my body. I know this moment, the one where there's no turning back.

"Guys, it's fine," I interject. "I'll just do it. I need to get better at math anyway, it's my worst subject." My mom whips toward me and fires in my direction.

"Don't protect him," she says, eyes cold as slate. "This dynamic between the three of us is so damn toxic. I can't believe you're siding with him yet again, Sarafina."

"Samia, she's not siding—" my dad attempts to interrupt.

"Don't interrupt me." Her voice is all icy fire, and I feel my body go numb, imagining my familiar iceberg, cold and strong and solid. My right hand reaches for my left wrist and I pinch, letting my nails dig in deep. I stare down at the marks I make in my flesh.

My dad capitulates, silently resting his chin on his hand and staring at her, eyebrows raised and eyes unblinking.

"Please continue," he says softly, waving her on.

"You two constantly gang up on me and I am tired of it. Sarafina, your disrespect is reprehensible. And Peter, you should know better." She's heaving, like a runner at the end of a race. "I'm done with this. I can't do it anymore." She twirls on her heels and leaves the room, retreating to her bedroom upstairs. Once she leaves, my tears begin to fall.

"I'm s-s-sorry I didn't m-mean to cause a f-f-fight," I stammer between sobs. My dad lifts me up and stands me in front of him, hugging me tightly.

"This isn't your fault, little one," he murmurs in my ear. My body shakes as I cry, and no matter how hard I try to hold on to that image of the impenetrable iceberg, my dad's warmth melts into me.

"It's okay, honey. You're okay," he repeats until I wipe my sleeve across my nose. Summer comes back and curls herself around my feet.

"I'll do it. I don't want to disappoint you. I'm really sorry." I sniffle, unzip my backpack, and toss the packet of extra credit onto the table. It makes a gentle thud when it lands, a thick stack of papers crisp and white.

"You're not disappointing me!" He surveys me, face bunched up in concern. "Your mom is probably right. We should take a break and revisit it in the next couple of days."

But I know, deep in my bones, that this is the right thing to do. I want to get better at math; I don't want to feel like I'm drowning every time I crack open my textbook. And above all, I don't want my parents to fight anymore. If I show them that I want to get ahead, not because of my dad but because I'm committed, then maybe they'll agree and the fight will be forgotten.

I lift my head and meet his eyes. "I'm positive."

He nods, expression unreadable, opens the packet, and we get to work.

By the end of the school year, a happy exhaustion seeps into me. I made it through my first year of rigorous classes with a new group of friends. I loved my classes and my teachers and, for the first time, felt intellectually pushed hard. It was a wonderful feeling, similar to how my legs feel after a long tennis practice, pumping full of adrenaline and endorphins and working hard to get stronger, better.

Each grade has its own graduation ceremony, and so for the final time that year, my parents and I drive to St. Stephen's as a family. They argue in the front seats about money, and I'm silent in the back. From the depths of my backpack I pull out *Ender's Game*, desperate to tune out this conversation and be transported out of this world into another, fantastic one.

When we park at the base of the hill, my mom tells me to leave the book behind, and when I stick it back into my bag, my heart rate jacks up once again. Together, the three of us walk up the hill in a staggered line, me in the front, my dad a couple of steps behind me, and my mom bringing up the rear. To a passing observer, I imagine we look like a random group of strangers making the trek together.

I expect little from the graduation ceremony; as a sixth grader, it's more of a performance than anything. The eighth graders have a real purpose for their ceremony as they graduate to Upper

School, but for us, it seems like little more than lip service for our efforts.

Lillian and I have chosen matching dresses and sit side by side in the pews of the chapel, giggling quietly together through the ceremony. Our striped gray-and-white dresses get caught in the cracked seagrass cord of the rush seats every time we move, and we play a game to see who can get stuck the least.

All of a sudden, someone says my name loudly and I look around, confused where the sound is coming from. My sixth-grade classmates are looking at me expectantly.

"Go!" Lillian pushes me a little, and I stumble out of my chair. I look up and see our minister and Mrs. Duren standing side by side at the altar, looking down on me. Mrs. Duren's hand is beckoning me to the stage, and I hear nothing but the dull roar of the audience's clapping. I focus on walking straight, one foot in front of the other, utterly confused why I'm being called up to the front. I keep my gaze trained on Mrs. Duren, who nods encouragingly.

"Congratulations on the Award for Excellence in Math, Sarafina," declares our minister to the entire room. Bolts of lightning shock ripple within me.

"What?" I hear myself saying out loud, and Mrs. Duren grips my hand into a handshake and whispers in my ear.

"I'm so proud of how you've shown up this year. You've demonstrated an extraordinary amount of work ethic and drive. You deserve this." She pulls away and smiles, handing me a math book as my award. I stand next to her, stunned, until the clapping dies down and she softly reminds me to return to my seat.

Lillian and Emilia are waiting for me, grinning ear to ear. They both have their own collection of book-shaped awards, and I sit between them in shock, staring down uncomprehendingly.

Me? The most excellent-in-math student in our entire class? I know I didn't have the top grades or the best scores, so how on earth did I earn this award? By the end of the ceremony, I've convinced myself that Mrs. Duren made a mistake.

My parents rush up to me as we exit the chapel, brushing past the crowd of students eager to officially begin their summer. My dad is grinning like a maniac and my mom's anger appears to be forgotten. She's still clapping, and they both go in for a proud hug.

I cast my anxieties to the side as the three of us embrace together for a long moment, one so unfamiliar that tears blur my vision. The look on my dad's face is enough to carry me through another eon of grueling hard work and packed schedules. It's enough to break my heart, his pride and love written clearly across his face.

He kisses my forehead and claps my back in another hug. My mom nods in the background, a moment of solidarity between the three of us. I grip my book award tightly and grin.

"Makes those long days of driving across the city and studying with you worth it!" my mom exclaims, smiling. I furrow my brow, annoyed that she would make my success about her, but my dad places his hand on my shoulder knowingly. I take a deep breath and for a moment realize I have a choice: spoil the afternoon or appease my mom. In a rare moment of self-control, I choose the latter.

"Yup," I grind out. "Thanks for everything, Mom."

Mollified, she points out, "I wish you'd thanked us from the altar when you were accepting the award. You didn't do this alone, you know." She walks primly in front of me as she talks, leading us back toward the car. My dad, sensing my discomfort, expertly moves the conversation along, and I try to ignore the frustration that replaced the pride I'd felt just moments before.

"Feels good, doesn't it?" my dad asks when we settle back into the car.

"What does?"

"Pride. Knowing you accomplished something hard, despite what other people might think or believe." The ignition rumbles as he pulls onto the road, and I sit there quietly, thinking about what he said. I remember Mrs. Douglass's words of doubt, and the wretched self-hatred I felt afterward.

"Yeah," I admit slowly. "It really does."

And the knowledge sinks into me, heavy and thick, how to maneuver my parents to get along with each other and love me: work hard and do well.

The monster deep inside me chomps on my insides, and I subconsciously cup my hands around my center. I'm too unaware of myself to recognize the signature anxiety pain. Instead, I look to my parents and watch my mom now laughing without reservation, for the first time in a long time, genuinely reaching for my dad's hand. I see the look on my dad's face as he turns to me, one of unadulterated love and pride for me, his successful daughter.

I know I will do anything to keep those looks forever.

PART II

PHASES

CHAPTER V

THE EMPTINESS OF SPACE

There are over 200 billion trillion stars floating in the cosmic abyss of our universe. These glowing orbs of fire and gas dot our night sky, shining brightly against a canvas of never-ending black. Each star has its own evolution story, perhaps with its very own solar system of planets and moons, orbiting the center of a foreign galaxy far, far away.

To understand the scale of stars, imagine that you're on a beach somewhere on Earth, your skin drinking in the sunlight and your eyes grazing the azure blue of the ocean. Bits of sand get caught in between your toes as you walk down to the water, your heels flapping against the sand. You reach down and grab a fistful, feeling the grains rub into your palm as you look out at the expanse of the ocean.

In that one handful, you've picked up around ten thousand grains of sand. When you multiply that by the number of sandy regions all over the globe, from beaches to deserts to sand dunes to ocean floors, you find that the Earth has seven quintillion grains of sand.

Unbelievably, seven quintillion is a fraction of the total number of stars in the universe—there are over 10^4, or 10,000 times, as many stars as there are grains of sand on Earth.

It is rather remarkable, then, that the night sky is dark.

Stars are born when darkness collapses on itself. Swirls of gas and dust succumb to gravity, coalescing to form dense stellar cores that form the hearts of stars. As more and more material falls onto the budding core, particles collide, jacking up the temperature until the pre-star, or protostar, is so hot that nuclear fusion ignites in the core, jump-starting its stellar heartbeat. At this point, the star becomes "alive."

When protostars transition into full-fledged stars, they shine. Particles slam together in the stellar core, careening and colliding into one another and emitting light as a by-product. The light streams all the way out of the sublayers of the star to the surface, where it breaks through and finally shines freely. Nuclear fusion, in short, makes stars shine.

Stars are huge. Our closest star is the sun, and it could fit nearly 1.3 million Earths inside. But in the cosmic sense of things, our sun is relatively average-size; some stars can be as large as one thousand times the size of the sun.

For the most part, stars are not quiet. They're violent, turbulent spheres of fire and gas that spew out tremendous cascades of charged particles and radiation. These eruptions, known as solar flares, corona mass ejections, and prominences, disturb the stellar surface, ejecting searing-hot matter millions of degrees Kelvin into the surrounding stellar environment.

One might think 200 billion trillion of these devastatingly bright orbs is a number so astoundingly high that they'd light up the universe.

But the universe is dark, cold, and overwhelmingly empty. It's a lonely place. Despite all the stuff scattered around—the stars that glow; the planets that orbit; comets, asteroids, and whatever else—the universe is mostly vacant.

It's difficult to fathom just how empty space is. The observable universe is about 13.7 billion light-years across. Numbers of this scale are beyond our understanding—perhaps even beyond the human ability to conceptualize scales this size, emptiness this vast.

But flecks of objects are scattered across this empty expanse. Their very existence provides a framework of scale for the universe. Thirteen billion light-years of nothingness ceases to truly mean anything until it is measured *relative to something else*. Those objects dotting the cosmos, though scattered across incomprehensible distances, give meaning to the emptiness stretching between them. Without them, there would be a whole lot of nothingness stretching everywhere, indefinitely.

It's difficult to spot an object billions of miles away. But if you look far and wide enough, you're bound to find something.

Glimmers of stars floating in the cosmic ocean are those beacons on the blackest of nights.

The week before eleventh grade is to start, my parents tell me that they're moving away—and leaving me at home. In a rare moment of solidarity, they're standing next to each other in our kitchen, elbows so close they're almost touching. I'm sitting at the dining table and warily look up from my copy of *Harry Potter and the Order of the Phoenix*, the pages now so worn they're falling out of the broken binding.

"This year is going to look a bit different from the rest," my mom is saying, worry furrowing her brow. "I'm sure you're aware how we've been struggling financially—"

"Not *that* much," my dad interrupts. A shadow darkens my mom's face, and she glares at him.

"Yes, that much," she snaps.

"We weren't even sure we'd be able to pay for St. Stephen's this year." He huffs, his ears turning a faint shade of pink.

"The point is, to stanch the bleeding"—she says this part viciously—"your dad and I have both accepted jobs." At first, I am unfazed, until I see her glance once more to my dad, heave a breath, and look back to me.

Suddenly, I am sure I don't want to hear the rest.

"I've accepted a job in Virginia working for the Defense Department."

I blink. Virginia?

She heaves another breath. "Your dad is moving to Houston to work for JPMorgan. You'll be living alone for the year."

A thick silence descends upon the room, at odds with the loud buzzing taking place in my ears. I look up at them blankly and then back down at my well-worn book, wishing I could disappear into the pages.

I can't comprehend what she is saying.

My mom places a comforting hand on my shoulder, but I slide away. I keep my voice monotone.

"I don't understand." I can't bring myself to say more.

My dad sighs and, visibly bearing the weight of all of our stress, meets my gaze. "It's been difficult finding a job in Austin, little one. Most banks I've been looking at are headquartered in New York or Houston, and given my specialty in energy and economics, Houston is a good fit. I'll be close enough to you that I'll be able to see you most weekends and continue taking you to tennis tournaments."

My stomach twists. *Tennis. Fucking* tennis. I snort under my breath and slam my book shut.

"Your day-to-day schedule won't change too much," my dad continues. "All you need to do is focus on school and tennis and try not to get too distracted by all of this noise."

"Right. Okay," I say dully. I push the chair back in an effort to get up, but my dad lifts a hand to signal me to stay.

"It's okay to feel overwhelmed," he says gently. "We'll still be here for you, just a little farther away."

My mom's eyes well with tears.

"I will miss you, my angel child," she says, voice breaking. From what feels like another world, the kettle starts whistling, breaking the moment. She bustles away to take it off the heat. For a brief moment, it's just me and my dad.

"What do you think, honey?" my dad asks me. He looks relieved to have the hardest part of the conversation behind us and sits down on one of the barstools to my left. I watch the meat of his hands pool across the granite as he lets his body sag against the counter.

Watching him like this—tired beyond exhaustion, cowed by this harsh reality—pulls at something deep inside of me, and my heart aches. I do not know what I think, what I feel, save that I can't bear to see him in pain. Like a magnet, I move to comfort him.

"Don't worry, Dad. I got this, I'll be fine," I say in a reassuring voice. He lifts his arms, hesitantly at first, and then bear-hugs me.

I can feel his gratitude radiating through the palms resting on my back.

"I know you will." His voice gently breaks. He moves away, but I reach for his hand.

"It's going to be okay," I say.

My hand barely fits around his, the many calluses from gripping my tennis racket scraping along his knuckles.

He heaves a breath and swivels his head to look at me. A grin lights his face, all teeth and eye wrinkles, and he winks, eyes warm blue as the midsummer sky.

"I know," he says. "And even if it isn't, I still have you!"

I ignore the terror rocketing through me when I smile back. Already, I am making a mental schedule of how to function without my parents. Cook, clean, do laundry, go to school, take the dogs on a walk, attend practice, study until my brain won't work anymore. Not once does the thought occur to me that this could be fun, that I could have boys over or throw parties. This is a *responsibility*, and I refuse to let my parents down. To make this harder on them when they are already doing so much.

Failure is not an option.

I just need to keep going.

The week before my mom moves to Virginia, I hear her whisper that she wants to die.

She's perched on her makeup stool in her bathroom, staring at herself in the mirror, and I'm hovering near the bathroom door, frozen. It's a room of silver mirrors, two sets running parallel along the extent of the room, one above the twin sinks and the other forming the closet doors. The mirrors reflect off each other, creating an infinite line of Samias that stare back at us. Lines of pain are etched into their foreheads and around their mouths, and when they blink, tears roll down their cheeks.

All of a sudden, she jumps up and kicks her half-filled suitcase. I spot a lime-green bathing suit lying on top of a pair of knee-high suede boots, their pointy heels knocking into her toiletry bag. My favorite gray cashmere sweater of hers is intertwined with bottles of pills. She grabs one seemingly at random and unscrews the white top, popping a pill like candy into her mouth. I watch her larynx bob as she swallows it dry.

Half of me wants to comfort her, but the other half—the part I'm ashamed of—desperately wants to leave. My body betrays my indecision, one foot halfway into her bedroom and the other stuck on the bathroom tile.

She reaches for my hand and I finally move toward her, allowing her fingers, long and slender like mine, to twine our hands together.

"You can come with me," she says in a hushed voice. "We can transfer you to a school in Virginia and you can live with me in my flat." The fantasy she paints dances across my eyes: falling autumn leaves and towering pines, mulled wine and thick blankets, columns of books and . . . and screaming, tears rolling down my cheeks, locked doors, and silent meals. My heart thumps unevenly—I am trapped.

She drops the shirt she's holding into the suitcase and straightens, turning toward me so that the infinite line of Samias are facing me like troops waiting to strike.

"Do you want to come to Virginia?"

The weight of a million Samia gazes slices into my chest, piercing like knives. It's an impossible choice, and I don't want to hurt her, to deepen those lines of pain already etched into her face.

My stomach bottoms out as the thoughts swirl—will she be

safe living alone in Virginia? Will I be responsible if something happens?

Still, a small voice deep inside me knows that I can't go with her—I don't *want* to go with her and leave my life behind.

I don't want to be responsible for her, for her pain.

"I need to stay," I whisper. I can barely hear my voice, so soft I'm not sure the words actually left my lips. But her face changes, and the pain is replaced by a cold mask of rage. She stops crying immediately.

"Of course. Why did I ask?" she says, her voice rough.

"Mom, I love you," I say. "I'm sorry."

"When I'm dead," she says, glaring at me, eyes wild, "you're going to regret not being kinder to me."

In the last week of August, my mom gets on a plane and flies to her new home. The following weekend, my dad slides into his Volkswagen and leaves me too.

As I walk through the empty house, something inside me, something worn and cold and utterly alone, breaks. In a way, the shattering is a comfort. With it comes a blissful emptiness, one that spreads all the way across my limbs and down my toes until everything is numb and I feel absolutely nothing at all.

It's not until I finally step foot in Mikan's observatory on the first day of school that I feel a glimmer of lightness again. The entrance is plastered with stickers: a black-and-white number 42 right above a large sticker emblazoned with THE DUDE ABIDES. Next to the door handle, a beautiful sticker with the night sky, a gigantic telescope, and the words *The Observatory* hints at what lies behind the door.

I've dreamed of this room for years. Since I first entered St. Stephen's in sixth grade, but in a way, I have always been reaching for this, even before I ever knew it existed. I've held tightly to the image of the observatory as though I were carrying around an old picture, one that I've stared at so often that it's stained and slightly frayed at the edges. A jolt of excitement shoots through my body as I heave the door open and step into another world.

The classroom resembles what I imagine a scientist's playroom might look like. It's a smorgasbord of science objects: There are lab desks covered in 3-D models of our solar system and books with images of the universe opened to random pages; three or four smaller telescopes stuck in the far-right corner of the room; a life-size cutout of Albert Einstein; cases of rocks with model planets resting on top; stacks of decades-old notebooks reaching all the way up to the ceiling; gigantic maps highlighted in yellow and orange; gas canisters, magnets, and four shades of lava lamps; a three-foot-tall sparkling lavender-and-periwinkle geode; a dozen portrait busts; diagrams of the sun and stars; cabinets filled with magnets and construction equipment; meter sticks; and nerdy jokes (*A neutron walks into a bar and orders a drink—the bartender gives it to him for no charge!*) papered across the walls. Dozens of old newspaper articles and magazine cutouts are tacked onto the wall next to the door with headlines like A SUPERNOVA SHEDS LIGHT ON DARK ENERGY and SOLVING THE MYSTERY OF THE MISSING NEU-TRINOS. Toy rockets stand next to several photos of Saturn, a large diagram of solar spectra, and a thirty-year-old slide projector.

It's part shrine, part dream laboratory. Rows upon rows of photos of scientists and science communicators line the walls, many of whom I've never seen or heard of. Albert Einstein and Carl Sagan claim the most frames, and Sagan takes up the entire mantel at the front of the room. My stomach drops half a notch as I realize there are only two, maybe three women up there; the rest are men.

The laboratory provides the foyer for the throne room: the observatory. Although our first class is about to start, I walk there dreamlike, as though magnetically pulled. It's to the right of the lab room, round and dimly lit, and at the center is the most massive telescope I've ever seen. The eyepiece rests at about my height, but the tube that extends from it is easily six feet tall and two feet wide. The room's roof was long ago replaced by a gigantic dome that rotates alongside the telescope. Next to my elbow is a computer plugged into various parts of the telescope. Red lights line the walls and blink hazily against the daylight streaming in from the open door at the front of the room.

It's unlike any science classroom I've ever been in.

Mikan himself looks like he belongs to the room. He's aged in a wizened way, with thinning salt-and-pepper hair and a combed silver beard. He's standing in front of the chalkboard at the front of the room, watching us take our seats with twinkling blue eyes not unlike my dad's. His mouth moves easily into a smile, and laugh lines etched around his eyes and lips give the impression that his favorite facial expression is a grin.

I take my place at the front of the room directly in his line of sight. My black-and-white-speckled composition notebook rests below my palm, and I imagine that it's waiting breathlessly to be filled. An airy excitement subsumes me, reflecting the magic of the room. A room so different from the empty ones that await me each night now that my parents have left.

"Welcome to astrophysics!" Mikan's voice booms loudly, surprisingly invigorated despite his obvious age. He walks slightly hunched, the very image of a mad scientist, to the slide projector and flips it on, plunging the room into darkness except for the images projected onto the screen.

"I'm so happy y'all are finally here! It's been a boring summer alone in this classroom, you know." He chuckles to himself, and it's so light and contagious that the students laugh too.

"My name is Mikan—not Dr. Mikan, just Mikan is fine—and I've been teaching for . . . thirty-seven years? Thirty-eight? I've lost track." His mouth breaks into a wide grin and we smile back at him, enchanted.

"Those old composition books"—he waves his hand to the corner of the room, at the stack of notebooks threatening to fall over—"are from previous classes that I've taught. Lesson one in this class is to keep your notebook and document it clearly; you'll thank me later."

I grab my pen and scratch into the first page, *Document clearly!*

"This class is going to be very different from your other classes here at St. Stephen's. I don't use a big thick textbook, and I'm not going to make you memorize all sorts of terms. This class is about curiosity and community. Not in the way that this place tends to use the word 'community'—you know, as a buzzword—but *real* community, where your peers are your colleagues, and we ask hard questions together." It's a speech unlike any I've heard from a teacher before. Even in science classes, I've been ordered to memorize and master. This raw, unbent curiosity and kinship gives me goosebumps. I subconsciously rub my arms and scoot to the edge of my seat.

"I design this class to get you interested in asking scientific questions," he continues. He flips the slide projector, and on the screen shines the most beautiful photo I've ever seen.

It's a star in the night sky, but with gorgeous shades of sapphire, sage, and rose surrounding it in an incandescent bubble of gas. It simultaneously looks as old as the universe and young as a newborn star. Hundreds of shimmering white dots—other stars and galaxies, I realize—encircle the bubble, wrapping it in a diamond velvet embrace. I stare and stare and stare, mesmerized. The lava lamps flicker in red and blue and Mikan's voice cuts cleanly through the room, his voice echoing the magic of the observatory.

"This is called the Crab Nebula. It's one of the many objects you'll learn about this year." I scratch the information as quickly as I can into my notebook, trying to get every single word.

"In this image, you can see the dead star, or white dwarf, in the very center of the bubble. Around the star is a bubble of gas showing the outflow of a supernova explosion."

Staring at the images, I'm enraptured. They make it easy to forget how empty I feel with my parents gone. How can I feel alone when each picture reflects an entire universe to be explored?

He shows us exploding stars and white dwarfs, simulations of black holes and gravitational waves, moons and planets, ring systems and meteors. Throughout it all, Mikan ogles alongside us. It's like we're discovering the whole universe for the first time together. He laughs when we laugh and gasps when we gasp. He often pauses between images to share a quick fact or story, and each one is better than the last, until by the end I'm struck mute in awe.

The forty-five-minute class passes so quickly that it feels over before it ever truly began. We blink blearily as he flips the lights back on, temporarily blinded by their fluorescent assault.

"So, are we in love with astrophysics yet?" he asks, grinning. We laugh and shout in confirmation.

He walks over to one of the lab tables and picks up a stack of papers. One by one, he hands them out.

"This is Albert Einstein's famous 1915 paper on general relativity. Don't get too caught up in the math or language, but I want to start exposing y'all to *real* scientific literature. Try to think about the big questions of this paper and why they're important."

I grab the paper quickly, skin flush with excitement, and stick it in the front of my notebook. Each of us thanks him as we exit the classroom, and he nods and waves kindly goodbye.

I squint hazily as the sunlight pricks my eyes painfully once I get outside. I pause for a moment before rushing to art history, my next class, and look back at the observatory. The dome shielding the telescope towers up toward the sky, a flat white against blue.

At this, the empty black oblivion I've found over the last few weeks since my parents left recedes slightly. Shards of light break through, and when I take a breath, I feel something other than emptiness: excitement.

The air gets colder. Christmas lights spring up around our neighborhood, twinkling in the dusk, and their light guides me on my early morning runs when it feels as though I'm the only person in the world. Leafy garlands dotted with bright-red ornaments wrap around patio banisters, and the Lexus SUVs that are the staple of the suburban mom sprout reindeer antlers and Rudolph-red noses.

Despite the fact that I love astrophysics, Mikan's class is still extraordinarily difficult. The math in our problem sets occupies most of my time, and when I'm stuck, doubt sinks in, momentarily transporting me back to my math class when the teacher observed, "Only the four boys sitting at the front of the classroom are capable of doing any real math. The rest of you are hopeless."

I can't picture what a quark is, or why infrared electromagnetic waves are important. I'd expected the material to be difficult, but it feels like no matter how much I stretch my mind, the physics is beyond me. I make mental notes of the questions I have but keep them to myself, sure that everyone already understands those topics.

I hate my brain for not understanding quickly enough. To offset how stupid I feel, I redouble my efforts. I make late-night trips to the grocery store to stock up on energy drinks and learn how to make myself coffee in my mom's old coffeemaker. The drinks gurgle in my stomach, acid churning from anxiety or caffeine or both.

I am often so tired that my diagrams seem to quiver on the page, vibrating in concert with the lines I draw across them. I stress-dream in physics, dreams in which I recite formula after formula. I awake panting and go through four sets of sheets, each one stained with salt and sweat.

My professors know of my parents' arrangement and are aware that I'm living alone; still, the work they assign is unrelenting. I learn how to forgo sleep in order to study late into the night—and to compartmentalize the all-consuming anxiety about my parents, about tennis, about exams. When I study, I use three highlighters: yellow for new terms; pink for concepts; and purple for methods of thinking, analytical processes that I've yet to master. Sometimes, I fall asleep at my desk, and they puddle together in a rainbow amalgam on my lap.

Sometimes, I think of Grannell's advice: *make good choices.* To become a computer programmer, did she struggle in math and science like I do? Or did it all come naturally to her, like one of her tangram pieces perfectly sliding into place?

I can do this, I resolve to myself. I can work long hours on impossible material, pushing myself until I physically can't push anymore. Anything to avoid failing, to avoid that dormant anxiety that lives ready to ignite in my body.

Somehow, the knowledge is reassuring. I will do everything I can, no matter the cost.

When my parents call, I tell them that I'm fine. "Fine" becomes my favorite word, nondescriptive enough that I'm not entirely lying. One evening, when my dad asks how school and tennis are going, I bite my tongue so hard that it bleeds.

I hate that I failed to keep my parents' marriage together—deep down, I think that if I had just been better, they would have stayed. So I redouble my efforts toward tennis and school. I think of myself as a robot, nothing more than an A+ and a tournament win, a strong backhand, a penetrating first serve, a shaky forehand, a solid-A English student, mediocre in science and hopeless at math.

The subject material of Mikan's astrophysics is all that is keeping me afloat. Nearly every class, after introducing difficult physics concepts and equations, he reminds us of the expansiveness of the universe, all of the other planets and worlds and moons, stars and suns, galaxies and galaxies, those pinpricks of light that make this universe livable.

And that perspective of how small we are in the expanse of the cosmos—how little our problems matter in the grand scheme of things—is my anchor. I let it soak in and form a wall against the tide of numbness that's seeped in since my parents left.

During the middle of November, I'm sitting in Mikan's class gazing up at a pie chart titled "The Composition of the Universe."

The chart's a large circle with three unmarked slices. Three terms are written on the bottom of the chart, *Dark Matter, Dark Energy,* and *Ordinary Matter*—the names of the slices of the pie.

I expected that the biggest one, about three-quarters of the pie, would be matched with "Ordinary Matter," but I'm fixated on the other two terms: "Dark Matter" and "Dark Energy." They remind me of highly sought-after ingredients in a potion, or forbidden spells from *Harry Potter* in Defense Against the Dark Arts.

"As we discuss the universe, it's important to define exactly what we're talking about. This chart summarizes the composition of the entire universe—or at least, what we know of it so far." Mikan winks and circles the smallest slice, a minuscule fraction that I have to squint to see, in bright-red marker.

"Look around the room. Look at each other." We do, meeting one another's gaze and giggling awkwardly to ourselves.

"Every single thing that you see right now . . . everything you *feel*, touch, taste, and smell makes up this tiny slice of the pie. Everything." He pauses and the room plunges into total silence. I stare at Mikan like he has three heads, certain I must have heard wrong.

"In short, everything that we know exists makes up this minuscule fraction of the universe. Everything else is unknown . . . waiting to be discovered." For a moment, his voice turns dreamlike, and I can feel the magic of his words washing over me in waves, one after the other with the unrelenting force of a thousand tides.

He smiles knowingly, all too aware of the impact of his words.

"Ordinary matter is a minority of the universe," he continues. "The evidence indicates that it makes up 5 percent of everything."

A science and math genius who frequently helps me with precalc abruptly raises his hand.

"How is that even possible? If everything we can possibly count and understand is just 5 percent of the universe, then what could the rest possibly be?" Heads bob, and I find myself nodding too.

Mikan dips his head in acknowledgment. "That's a good question, and something that's baffled scientists for decades. How can the universe be full of stuff that we can't detect?"

He underlines the remaining two terms, "Dark Energy" and "Dark Matter."

"Everything else, the other 95 percent, falls into these two other buckets: dark matter and dark energy." The words reverberate through the silent classroom like a gong, heavy and metallic. I imagine whorls the color of dusk sweeping through the room, dancing in between the desks, swollen with the weight of a million invisible worlds.

"They're a total mystery. We can't see, or touch, or feel them; they're invisible and, in many ways, undetectable."

Reeling, I copy the chart into my lab book and write the words slowly, trying desperately to wrap my mind around each letter. How is it possible that we know so little about our home, our universe?

I have so many questions. How do astronomers detect dark matter and dark energy if they're invisible? What do they tell us about the rest of the universe? Why are they both called "dark"? Is dark matter like normal matter by inherently being related to energy?

I wonder if I'm stupid for wondering about these questions, whether I'm supposed to already know the answers. I flip back through my composition notebook to see if the terms were mentioned during any of our other classes. I find nothing, and my confusion builds.

Mikan continues, "In many ways, astronomy is the science of discovery. Just as ancient civilizations explored new lands, we explore the universe. The cosmos is the final frontier." At this there is a smattering of laughs, and Mikan grins again.

"Some *Star Trek* fans . . . good, good, just want to make sure you're paying attention. Don't forget that you're supposed to be halfway through *The Physics of Star Trek*, by the way." A couple of us smile, and the spell that had struck the class mute breaks. People begin moving in their seats again; some stretching, some fiddling with their notebooks, and some going so far as to swipe open their phones. Mikan clears his throat, and the movement slides once again to an abrupt stop.

"For those of you who aren't in love with astronomy just yet, you'll find that the passion for discovery is ubiquitous in all of science. Contrary to what you might think, science isn't boring or stuffy. It's not just solving math problems or plowing through derivations." He points to Jenny, who is sitting three seats down from me, and she blinks her black bangs out of her eyes. "Speaking of which, I think you owe me a derivation or two?"

She motions to apologize, but he stops her.

"It's okay, it's okay! It was just a small reminder . . . take your time, I know y'all have enough on your plate as it is." He walks toward the chalkboard, where the pie chart projection shines against the black slate.

My questions swirl in my mind, but I try to listen to what he is saying.

"Remember the story I told y'all about the Grand Canyon?" At the beginning of the semester, Mikan told us how he trekked the Grand Canyon in three days, through moonless nights and in temperatures just above freezing. I remember because I looked up pictures of the Grand Canyon afterward and decided that one day, I want to hike under a starlit sky in the middle of nowhere too.

"That trip only happened because I was doing science. Some of my favorite memories—some of which I've shared with y'all— only happened because I get to do science. I've gotten to camp and hike in some of the most isolated, hard-to-reach places. Science is about getting your hands dirty. It's about playing, and curiosity.

"We're already five minutes past the bell—" Lillian shoots up in her seat, eyes wide, and scrambles to stuff her notebook and pen into her bag. "I just want to leave you with this, and then I'll let you go." She pauses, slowly setting it back down on the desk. Mikan nods kindly, switching off the slide projector and flicking the rest of the fluorescent lights back on. I blink once, then twice.

Once we are still, Mikan says quietly, "This is the most important lesson I am teaching you this year. It goes far beyond any exam or final, so listen carefully." I strain forward in my seat, setting my pen down so softly that I don't hear it hit the table.

"Science isn't about how many problems you can solve or equations you can memorize. It's about the thrill of exploration. It's a desire—and a decision—to endeavor to understand the unknown." He points to the cardboard cutout of Einstein, and then the framed photo of Sagan.

"Both of these people were immensely curious about the world. They sought to understand our existence. *That's* what makes a scientist."

I picture what I thought I knew about scientists—men typically, with disheveled salt-and-pepper hair, who are white and old and a little wrinkly and don enormous glasses and starchy lab coats behind a locked door, but Mikan's description—adventurous, playful, dirty—doesn't fit into that vision. His scientists are climbing mountains and floating in space, digging into the depths of the world and swimming with the sharks. I see myself in them; we are not necessarily brilliant but *curious*. Together, we form our own band of question-askers, "scientists," and while that room of old men in white coats is still locked, it doesn't matter, for we get the whole universe and they get a blackboard and broken nubs of chalk.

That night, while I'm poring over my problem set, I resolve that the only way to improve at astrophysics is to ask questions. The questions from class today are lit like embers in my mind, and I want to understand their answers.

I'm tired of slogging through this material alone, too afraid to ask questions for fear of looking stupid. I flip back through my class notes and problem sets, writing down each question that pops into my mind. I'm sure that some of them betray fundamental gaps in my knowledge, which I know will be embarrassing, but at least I'll finally get to learn the answers.

Hands steady, I type an email to Mikan, asking to meet. Almost immediately, he responds, requesting that I swing by tomorrow. I grasp my pen tightly and resume forming my list of questions, waves of trepidation cresting into me.

When I walk into the observatory the following morning during my free period, bleary-eyed and heart racing, I place my books down.

The room is entirely empty. Without the bustle of class or intrusive pen-scratching on paper, it's quiet—peaceful. The door to the telescope is slightly ajar, and when I spot the midnight-black star charts, I pause. My page of notes slips through my fingertips and falls to the table, forgotten.

I walk, dreamlike, over to the telescope. Pinpricks of light perforate the dark, Earthen stars that glitter above me. The room smells of parchment and dust and moonlight. A star chart lies forgotten in the corner; someone has penned a series of target constellations; someone else has left behind a notebook of coordinate systems—the room bursts with the prospect of discovery.

My heart stills and my palms cool, and as I stand there in the silence, a wave of calm rolls through me.

I feel closer to the sky in here. Although the dome is closed, I imagine it rotating counterclockwise, ticking peg by peg until it halts just above the telescope mirror. I picture the dome door cranking down like a window until a pocket of sky finally becomes visible and the cosmos shines through.

In my vision, I hover, suspended in this purgatory, the Earth pulling me down and the sky tugging me upward, up, up, up.

This must be my heaven, I think to myself.

A place unencumbered by the shackles of the ground, so close to the sky you can nearly taste it. I take flight across the sea, the sharp scent of sea salt winding to my nostrils as I float ever higher, past the ridge of clouds and into the stars.

I might never reach it, I know, but if this is as close as I get, then I'll consider myself lucky.

My dreams don't chase me here.

In this non-place, I realize, I am free.

"How can I help you, Sarafina?" Mikan's voice cuts through my reverie, grounding me. I look down and see my feet planted firmly on the floor, just one step from the telescope. If I'd taken half a step more, I would've walked straight into the eyepiece dangling ninety degrees from the optical tube.

I blink and shake my head and the vision fades, but my dread remains at bay. Instead, that feeling of suspension—of peace—has taken its place.

Mikan is leaning on the doorframe, arms crossed casually. I blush, mortified that I was caught gazing up at a closed dome, and hurry back to the desks.

"I was hoping I could ask you a few questions," I say. When I pick up my list, I brace myself, anticipating my mind filling with an anxious fog. I wait, knuckles white, until I realize it's not happening.

My mind is clear. My favorite questions jump out at me from the pages, and they hang frictionless in my mind. A small laugh escapes me, for I didn't know until this moment that I even had favorites.

Mikan eases into a chair facing me. The hinges squeak as he reclines. He sips his coffee and looks at me, waiting for me to begin.

I've never sat with a teacher like this, like equals. My stomach still doesn't drop, and I realize that his posture sets me at ease, almost as though he wants to learn from me just as I want to learn from him.

We work through the first four questions, taking short breaks for him to demo general relativity (a bowling ball placed on a personal trampoline) and the expanding universe (a balloon with dots drawn all over it).

With every question I ask, he takes his time responding, explaining concepts so carefully, with so much excitement, that I think he might be enjoying this as much as me.

Not one single time does he make me feel stupid. When I ask how astronomers detect dark matter, he even says that I'm asking a good question.

This is the science he was talking about, I am realizing. This is playing, learning how to ask questions, indulging my curiosity.

It all leaves me breathless.

I love it.

It is when I look up at the clock after my fifth question, shocked to find that we have only six minutes left and we've not worked through even a quarter of my list, that Mikan pauses, settles back into his chair, and scratches his beard.

"A couple of decades ago," he says thoughtfully, "I taught a pair of brothers. They were great students, and their father was a professional astrophysicist at the University of Texas."

I jerk straight up, pulled upward like a marionette dangling in wait.

"It might be good for you to go down to campus and chat with him. He's a super nice guy." He smiles, then continues, "His name is Dr. Craig Wheeler. He works on supernovae and indirectly on dark energy."

I gasp softly and my heart starts racing. Still smiling, he says, "I'll make the introduction and you can take it from there. I think you'll enjoy learning from him."

"Th-Thank you," I stutter. "This is so kind, I would love to meet him."

He nods, blue eyes twinkling.

"Of course. This is all you—you earned it."

My heart warms, and I gather my papers in quiet shock. When I finally emerge from the observatory, I pause outside the building. It's a cold, dreary day, heavy with clouds and a light mist, but despite the weather I am so light I could float away.

The meeting went far better than I could have dreamed, and I let the realization sink into me. Mikan doesn't think I'm stupid or incapable of doing astrophysics. He didn't judge me on my questions; in fact, he loved them.

I stand still, letting the mist soak through my jacket, and think of that astronomer from fifth grade, the one at science camp who said that astronomy wasn't for me.

He's wrong, I think to myself.

A light breeze caresses my face, as though in confirmation. I tilt my face upward to the sky.

"I *can* do this," I whisper. As I walk away, something intangible gently tugs at me, as though pulling me back to the observatory, to that place that feels like home.

CHAPTER VI

FAMILIES OF STARS

Stars, like us, are born in nurseries—albeit nurseries of a different sort. Within *giant molecular clouds*, the stellar nurseries of the universe composed of gas and dust, gravity compresses overly dense regions, forcing them into collapse. From each fragmented, collapsed region, stellar seeds called protostars are formed.

Anywhere between 10,000 and 10 million protostars the mass of our sun are formed within the same giant molecular cloud. As they collapse, their internal temperatures and pressures rise, slowing down their fall until the star's inward force of gravity and outward force of pressure are balanced in *hydrostatic equilibrium.* At this point, the star's core becomes hot and stable enough to ignite nuclear fusion, the star's life source.

Deep within the stellar core, elements ram into one another. The pressure released by these collisions counteracts the inward force of gravity, stabilizing the star and allowing it to evolve. Collisions are stellar fireworks; they release photons as starlight, which streams outward all the way through the stellar surface and out into the cosmos.

Stars born from the same giant molecular cloud share the same stellar DNA in the form of elements dispersed throughout the cloud. Their composition shapes how they grow, evolve, and die. These groupings of stars born within the same family are known as star clusters. The oldest stars in the universe are in *globular* clusters and, for the most part, remain there until they die. Younger stars, on the other hand, often escape, thanks to being more loosely bound within their *open* cluster.

More than half of the stars in the universe are born with sibling stars, stellar companions that are gravitationally bound. They orbit one another, forming binary—or multiple—systems. As a symbiotic system, they shape one another's evolution and fate. Sometimes, gravity pulls them together, binding them into a merger. Others remain orbiting one another, gravitationally bound but never merging, until they die or are pulled into yet another stellar interaction by a nearby passing star.

Stars are shaped by their surroundings. Some stay with their stellar families, influencing one another until their deaths. Others roam the universe, creating new star systems as they travel through the abyss of space. And still others, the lonely ones, never form new systems, forced to roam alone until they die.

In the middle of January, when the air is thin and the sky is gray, my mom calls to say that she's quit her job and is moving back home.

"I hate it here," she cries over the phone, voice breaking.

Before I know it, just a few days later I'm picking her up from the airport. She looks the same—perhaps with a slightly stiffer upper lip and darker bags under her eyes—but she carries herself like no time has passed: tall, despite being five-foot-three; elegant, one heel placed directly in front of the other; and languid, like her favorite Julia Roberts in *Pretty Woman*.

She talks as we drive home, but all I hear is the grind of the gearshift and a faint gnatlike buzzing. A swell of emotions rises within me: anger that she's back, shame at my anger, resentment that I was left alone at all. But they're welded together, indivisible, and the mess of hurt rises so much that my skin feels like it's been set ablaze.

I grab my right hand with my left.

Pinch, pinch.

Harder. More nails, dig in deeper.

Pinch, pinch, pinch.

Harder, goddammit.

PINCH!!

Fuck.

I draw blood, and it pools in tiny beads by my knuckle. It's slight, so small I have to squint to see it. But still. Fuck.

I casually run my hand along my jeans and rub it off on my leg. Thankfully, the blood barely stains the stitching. I heave a sigh and my mom swivels around. I keep staring forward, and finally she turns back too.

When we pull up to the house, the flesh in between my thumb and forefinger is scratched red and numb but no longer bleeding. I tug my sleeves down all the way past my fingertips and wheel her suitcase into the living room.

"I'll drop you off at school in the morning," she says, "and then I'm going to run some errands. Do you need anything from the grocery store?"

"Crap, I'm sorry," I say. "I completely forgot to tell you. Tomorrow, I have to go to the University of Texas for a meeting."

"What? What about school?" She drops her bags on the counter, flustered.

"I'm sorry, I completely forgot." I pick at my nails, terrified of what she might say next.

She pauses, and her face relaxes a fraction, and then she asks, "What is the meeting for?"

"Mikan knows how much I love astro, and he's introducing me to a faculty member he knows. Basically, I just get to learn more about what his day-to-day is like, what kind of research he does. That sort of thing. I have a list of fifteen questions I want to ask him." I rush over to the kitchen table, where my stack of books lies, and brandish a sheet of paper with a typed bullet-point list.

She takes it, looks it over, and her face brightens.

"This is fine, angel child. I'll take you." I blink, a little surprised, but her expression is clear and warm, and I can tell that she means it.

My stomach rights itself, but part of me is still on guard, waiting for the other shoe to drop.

"Great, thanks so much, Mom. I really appreciate it," I say tentatively. "The meeting is at noon, so want to aim to leave around eleven?"

"Sure, that's fine." She collects the bags she dropped and walks toward the stairs. I bend down to pet Summer and pull her in tight, burying my face in her fur. When I hear the suitcase rolling away, I look above her ears.

"I am happy that you're home, you know," I say quietly. She bends down and kisses me on the cheek.

"*Shukran, habibti.* I'm happy to be home too. More than you know."

When I wake on Tuesday, January 25, the unexpected sound of the (English) television fighting over a blaring (Arabic) radio greets me, guttural and harsh *kh's* cresting against a plain, Texas twang. Mornings are no time for television, my parents have told me. I blink several times, clearing the sleep from my eyes.

When I get downstairs, television pundits are chattering to an empty living room. Images flash across the screen: swarms of people flooding Tahrir Square; crinkled red-white-and-black flags; the rotunda of Cairo University; bright flashes of light and thick clouds of heavy smoke. The towering Pyramids of Giza.

My mom is hunched over and peering through a pair of thick reading glasses, flipping through an old leather book of phone numbers. Her normally well-organized desk is cluttered with stacks of papers and an overturned inkwell, which is dripping undetected onto the carpet. The volume dial on the radio is turned all the way to the right.

"Mom?" I ask, beginning to panic. "What's happening?"

She looks up, startled.

"Oh, hi, angel child. I didn't see you there," she mutters. She's disheveled—curls askew and kohl smudged beneath her eyes.

Ears straining, I listen intently to the radio.

Years ago, when I was first learning to write, my mom tried to teach me Arabic. She bought me a book of the alphabet and we spelled my name together, sounding it out: S-a-r-a-f-ee-n-ah, سارافينا.

The letters jumbled together in my mouth and my mom trilled a laugh, *habibti hawil maratan 'ukhraa*; my love, try again.

When I was older, the language stopped being beautiful. In public, *yallah* meant pointed stares, *sabaah alkhayr* meant hissed insults.

Sand nigger. Camel jockey. Terrorist.

Min fadlik tawaqaf ean qawl dhalik. Please stop saying that, I begged my mom.

We didn't speak Arabic together after that.

The words from the radio ring violently in my ears. I understand none of them, those early Arabic lessons forgotten.

With a regretful shake of the head, my mom motions me over to her computer. I spot the orange-white cursive logo of Al Jazeera in the upper-left corner. In large block letters, it reads: EGYPT PROTESTERS CLASH WITH POLICE.

"Revolution has come to Egypt," she declares.

Eyes wide, I fall into the seat across from her. For once, the sticky tension that fills the space between us evaporates, momentarily forgotten.

"Thousands of Egyptians began protesting the president, Hosni Mubarak, this morning. They're demonstrating all over the country, but the epicenter is in Cairo. I'm trying to get in touch with Susu and Lubna," she says, voice breaking, referring to her cousins.

I can't quite picture it, the Egypt I know and have seen with my own two eyes, now so different in the news, the images from the television searing my mind. From across the world, the events feel faraway, but these are our *family*, in our *home*.

Most of my mom's extended family is still in Egypt. As much as she loves her home country, she chafes at the restrictions women face there—what we're required to wear, how we need to speak, what activities we're allowed to do. One trip, when I was just over two years old, we visited our family and she was detained by military police for carrying me across city lines without her husband. They called my dad, who was back in the States, to obtain his explicit permission.

While Egypt will always be her home, it's also her captor, and as much as she loves it, I don't think she ever wants to live there again. Still, it's home.

She looks back down at her notebook.

"Ah! Found it." She skims her finger along a line and picks up her cellphone.

Her attention remains fixed on the screen as a voice comes through the phone. "You are dialing an international number. To continue, press 1."

She presses it.

"*Marhaba, assalaam alykom.*" A woman's lilting voice floats through the ether. My auntie Susu. My mom exhales an audible sigh of relief and shoos me out, grabbing the phone to her ear and turning back toward the computer screen.

Slowly, I return to the kitchen to make breakfast. I put it together mechanically: two scrambled egg whites, two pieces of seed toast, one cupful of strawberries. No stems, no cracked shells.

A pundit, blond and spray-tanned, is yelling on the screen next to a picture of Osama bin Laden, *Infidel Leader.* Sighing, I reach for the remote and press the red button. It blinks off.

A learned sense of dread descends upon me, one I'm now familiar with since the Twin Towers fell. With the Egyptian Revolution coverage is bound to come an increase in racism toward Arabs, an increase in hate crimes, an increase in ways I feel *Other.*

The notes I was up studying late last night lie scattered on the table.

1. **What is the current status of the rapidly expanding universe?**

2. **Is there any evidence other than Type 1A supernovae?**

3. **Is there any more progress on the quantum-gravity connection?**

As my stomach sinks, the realization dawns on me that in just a couple of hours I'll be talking with an astronomer, one of the most important meetings of my life, and yet my mind—and my heart—is not with my books but with Egypt.

A newfound respect for my mom washes over me. I have no idea how she managed to balance her new American life with her Egyptian one. Since she moved to the States, Egypt has undergone assassinations, war, chaos, and every time she was forced to bear her grief from seven thousand miles away.

My mom emerges, eyes shining, from the office. She must have washed her face, the kohl rimming her eyes fading into the deep brown folds of her skin.

"I was able to talk to Susu and Lubna. Everyone's okay, thank God." Relief sweeps through me, until she says, "But the family needs to be very, very careful."

She moves with jerky motions, stress radiating from her body, and suddenly I'm glimpsing her in a new light. How brave she must be, to honor her own independence and fight for her education in a foreign country, in a foreign language. How much guilt she must be carrying, to not be in Egypt standing alongside her family. The overwhelming grief of being so far away from home and not knowing if they're safe.

It's only a flicker of understanding who she is, but it makes me wonder if her rage and despair are moored in something I can relate to more than I thought.

"But they're fine now, right?" Images from the television swirl in my mind of unnamed people swarming the streets, faces pinched and angry. The military police advancing upon them, brandishing weapons, and the fears bite in—will they stay safe?

"Yes." And then quietly, so softly I can barely hear, she says, "For now."

From across the room, she bends her neck to gather her things, bouncing curls shielding her face. She swipes a finger under her eyes, ridding them of smeared kohl, and heaves a sigh before turning around and looking at me.

"Ready?" She says it forcefully, as though convincing herself to part from her desk and get into the car.

Something about this moment feels wrong. The realization rockets through me that I should stay with her, that this is far more important than my studies. This feels like a moment frozen in time, one marked with fear and trepidation that we should share together, two Egyptians on the wrong side of the world.

But she strides across the room, head held high, and opens the door to the garage, and I follow in her steps, clutching my notes against my chest. We each slide woodenly into the car, one of us on each side, mirror images of each other, and silently drive into the morning.

When I look up at UT's physics and astronomy building, RLM, from outside the front door, coherent thoughts of Egypt begin to slip away and are replaced by a deafening anxious roar. The building looks joyless—twenty stories reach into the sky, much taller than any of the surrounding buildings, a solid heft of dull red brick. The cornerstones, inscribed with the names of long-forgotten scientists, are many decades old, their edges chipped and cracking. Inside, I imagine, are brilliant minds tackling impossible problems.

A heavy weight settles into me, tempering any excitement I might feel—I can't possibly belong here.

An icy wind blows as I reach for the door handle, and I tug my jacket closer across my chest. Mechanically, I walk to the elevator, one foot in front of the other.

With every step comes another anxious thought, a swirling mess of worry.

Is my Egyptian family safe? Are the ones protesting getting caught . . . beaten . . . tortured . . . or worse? Will it ever be safe enough for me to go back?

Scientists and students pass by, shoulders brushing against mine. Do they see through me—that I don't belong here?

Desperately, I begin to silently count to calm myself: nine steps to the elevator, and nine is three threes, three large square tiles I step on top of. Threes are solid and prime. They're my favorite.

Two more steps to walk inside, then the doors close and seventeen floors pass by as I rocket into the sky.

A button dings, and the doors yawn open to a long corridor. Framed photos of professors and research groups line the hallway. Most have shiny, milk-white skin that glimmers in the late-morning light, many with thinning hair and heavy-framed glasses, some with goatees or mustaches or long, unkempt beards. Two women hang in the line, but the rest are men, their collared shirts stiff and colorless.

My boots clack against the tiles. When I reach the end of the corridor, a gold-lined plaque hangs next to the room on the left.

DR. J. CRAIG WHEELER
PROFESSOR OF ASTRONOMY
SAMUEL T. AND FERN YANAGISAWA REGENTS PROFESSOR
OF ASTRONOMY

My mom's drawn, kohl-smeared face from this morning wavers in my vision. *Terrorist,* I hear, in the sharp, nasal voice of the blond television pundit. My heart is beating so fast I swear I can hear it.

Count, I remind myself. Eighteen inhales and eighteen exhales, two sets of nine, which is six sets of three.

I heave a last breath and knock once.

"Come in," says a voice.

I push the intrusive thoughts far, far away, into a deep corner of my mind, where they're stuffed in a box to which I've thrown away the key.

The door swings open and my feet carry me into a moderately sized room. Wide windowpanes span two of the walls, yielding unobstructed views of the famous University of Texas tower and the rolling Hill Country just beyond. Two tiers of bookshelves, each ten feet tall, line the wall directly in front of me, stacked with fraying books and model telescopes and golden figurines. I spot royal purple and black spines, with titles like *Electrodynamics* and *Modern Cosmology* and *High Energy Astrophysics* written in cursive white letters.

To my left is a blackboard, and the outline of a bald figure playing the trumpet is sketched in the upper-left corner, so intricately shaded that I can almost hear the buzzing of its horn. Equations fill the rest of the board, haphazard and messily drawn, some half-erased in dispersed clouds of chalk and others redrawn over and over again.

Dr. Wheeler swivels in his chair and faces me. He's thin and spry, with a full handlebar mustache and wisps of silver hair that droop across his forehead. Clear, bright-blue eyes widen behind a pair of spectacles when he grins.

"Welcome," Dr. Wheeler says, "to RLM."

We shake hands, and I thank him for taking the time to meet with me.

"Of course!" He smiles. "Mikan is an old friend and colleague, and I'm happy to chat with interested students."

I tug out my list from my backpack and rest the papers on my lap. Seven pages, each one lined with ten bullet points perfectly spaced apart, ordered from early universe to present. He raises his eyebrow at the packet.

"Sorry," I warn, "I have a lot of questions."

Eyes crinkling into a smile, he says, "Don't be sorry for being curious." And with that, he motions for me to begin.

My stomach roils once more in anticipation, and the equations on the blackboard blur. My brain tells the muscles in my face to move, to say something, but my mouth remains closed, dry, and silent.

And despite my best efforts to compartmentalize the news from Egypt, images from this morning begin to flash in my mind's eye. I see my cousins Mahmoud and Yehia, Hussein and Hasan, hoisting signs into the air in Tahrir Square. Rubble litters the streets and piles up in corners of alleyways as crowds rush by or take shelter behind makeshift barricades. Others gallop through the streets on camels, shouting in Arabic through plastic goggles and heavy neoprene gas masks. Protesters stand like kings atop burnt, decrepit cars, surveying a broken empire as firebombs paint the air bright yellow and palm trees burst into flames.

The sky rains smoke and stones.

Moments pass and Dr. Wheeler is still looking at me, now with concern furrowing his brow. Acid surges in my gut.

"I-I'm sorry," I blurt out. "It's been a weird morning." I pinch my left palm and feel a sharp pain, a reminder that I am here, sitting in this room, my feet pressing into the floor of a tower in the middle of downtown Austin, and I am across the planet from a revolution unfolding on another continent.

I pinch my right palm now, desperate to jump-start my brain.

He raises an eyebrow, waiting, and in that moment, I know I have to choose. Mind swirling, I heave a breath, and make the decision then to be open with him, to not bottle everything up like I always do, because when things get too unbearable and the bottle becomes too full, I collapse, and I cannot collapse here, not in this place, not in front of him.

"I'm Egyptian—well, half-Egyptian, and my family—" I begin, but his expression clears and he raises a hand to stop me.

"Oh God, Sarafina, I had no idea. Are they okay?"

"I think so. They're in Cairo, so they're in the middle of it."

He pauses, surveying me, and then says in a kind voice, "What an extraordinary moment to meet you." Nothing about his demeanor looks forced or judgmental; he reminds me of my dad, all smiles and understanding.

"I've been following the news in Egypt all morning," he continues. "We can take our time going through whatever questions you might have, but I understand that this is likely a complicated moment for you. There's no pressure here."

Blinking back tears, I look back down at my pages of notes and swallow a sigh of relief. The pink and yellow highlighted lines look sharper, cast in high-res.

With his permission, with his *validation*, I'm able to hold both truths in my mind: the Egyptian Revolution and the astronomy in front of me, and when I acknowledge they're both here, both important, they no longer fight for space in my mind. They can coexist beside each other, gently swaying together side by side.

"Thank you," I say, and he nods. "Okay, first question. The books I'm reading have said that the universe is 'flat.' But what does it mean to be in a flat universe? Is there a way to physically understand flatness?"

He looks at me for a moment longer and then leans back into the seat of his chair. A smile quirks his lips.

"Whew, so we're just jumping right in off the deep end." He chuckles, then says, "That is one great—and difficult—question."

"It is?" I accidentally ask out loud.

"Absolutely. It's extremely difficult to picture non-Euclidian geometry." I nod like I know what that means, but he shakes his head and huffs a laugh.

"Sorry," he grimaces. "What I mean by that is geometry that's different from what we experience every day. Here, let me show you."

He rises and walks to the blackboard. He picks up a long piece of chalk, sketches a triangle, and labels each of the three vertices: *Laser 1, Laser 2, Laser 3.*

"Imagine that each of the vertices are spaceships flying around in space. Each one shines a laser at another, so that they form a triangle of beams."

"Sure." I nod along.

"In 'normal' geometry, what we call 'Euclidean,' the triangle's angles always add up to 180 degrees." To the right of the triangle, he draws a sphere, dotting a line from left to right across the center where the equator might be.

"Imagine this sphere is the Earth. If you were to draw a triangle from the north hemisphere to the south"—he draws one across the equator—"it would look like this." The triangle is bloated with fat angles, no longer the sharp one of the spaceships.

"Now, here's the weird part." He turns to me, eyes bright. "The angles no longer add up to 180 degrees. They're now anywhere from 180 to 540." I gape at him in disbelief, and he nods.

"I know. It's bizarre. Tying it back to your original question; when we say that our universe is 'flat,' we're saying that it follows Euclidean geometry. A triangle drawn in space will always be 180 degrees."

"Oh! That makes sense, I think." I copy the shapes into my lab book while Dr. Wheeler returns to his chair.

"Astrophysics can be difficult because we have to picture complex ideas in our minds. We can't touch many of the things that we study; they're either too far away or too abstract. We have to come up with clever ways to ask questions and even cleverer ways to answer them." I write everything he says, word by word, into my notebook, and he waits patiently for me to finish.

As I scribble, my mind flashes yet again to Egypt, so far away that the Revolution is two-dimensional images blinking across a television screen, too abstract to fully understand, and it feels just as unreachable as the vast distances in the universe.

Yet my love for both is unchanged, poignant enough to make my eyes water and my heart hurt.

An hour passes, then two, and as the clock chimes three I jolt up from my notes. When he speaks with animation, undiluted joy lighting his face, he reminds me of Mikan. With both of them, there is no judgment, no shame.

We've covered dark energy and supernovae, the Casimir effect and subhaloes, MACHOs and WIMPs, dwarf galaxies and the Big Bang. My brain tingles it is so full, like it's been stretched every which way and is stronger for it.

I pack up my pens and notes, filing them safely into my backpack, and heave a sigh of relief. Despite meeting with an actual astronomer for *hours*, I'm more settled than I was when I walked into the building this morning.

When I stand up, I say, "It was a pleasure meeting you, Dr. Wheeler."

"It was lovely meeting you too. Thank you for all the wonderful questions," he says, and then he adds, "You know, science is as much about asking the right question as it is about getting to the right answer."

I nod, but to my surprise, he continues: "Based on today, I think you'll make an excellent scientist."

I don't remember shaking hands goodbye, or taking the elevator down to the ground floor, or walking three blocks east to the parking garage on San Jacinto, where my mom picks me up. His words ring in my mind as loud as the clock tower bell clanging at the center of campus.

"How is the family?" is the first question I ask my mom as we merge onto Lamar and cut toward Mopac.

"Everyone's fine, angel child," she says, "but I wish I was in the streets with them." My mom, calm for once, looks over at me. "How was your meeting?"

"It went well," I say, blushing, and leave it at that. But in my mind, I turn Dr. Wheeler's words over and over: *You will make an excellent scientist.*

The day's events swirl together, and I cannot believe that just hours ago we got the news that my family was in the middle of a revolution, and then I met a real astrophysicist, and he taught me extraordinary, magical things about our universe.

He said I will be a good scientist.

When I finally fall asleep later that night, it is to the thought that yes, I am indeed on the other side of the world, but really, we are together in the cosmic sense of things; we are all floating on a world dangling in the sunbeam of a remote corner of the universe, hurtling through space at 2.2 million kilometers per hour, into the yawning abyss of the unknown.

Together.

CHAPTER VII

EXPLODING STARS

In a never-ending sea of black, beacons of starlight tether us to the cosmos. But it is when stars cataclysmically explode that the brightest lights of our universe get lit.

While stars about the size of the sun fizzle out into *white dwarfs* at the ends of their lives, fading into darkness as empty shells of themselves, more massive stars undergo a different evolution—one that ends with their explosion.

Stars are like onions. As they age, collisions within the core create heavier and heavier elements. Lighter elements from earlier reactions get pushed out to the star's outermost layers, the stellar envelope, forming an onionlike shell of elements that encircles the core.

The motto "Live Fast, Die Young" applies to stars just as it does to humans. The bigger the star, the hotter, and thus faster, it fuses, hastening the end of its life. But extremely high temperatures allow massive stars to fuse heavy elements that lighter stars cannot, from carbon to oxygen to silicon and sulfur, until they are so hot and the final element, iron, so heavy, that they can no longer jack up temperatures high enough to fuse any longer.

For stars, iron is anathema. Elements heavier than iron are *endothermic*, meaning that they absorb nuclear energy instead of releasing it. The high temperatures and strong pressures that once maintained the structure of the star against the inward pull of gravity cannot get high enough as elements begin absorbing more energy than they can possibly release.

Instead of successfully fusing the now-iron core, previously fused lighter elements surrounding the core reignite fusion thanks to the star's now-obscene temperatures, which climb into the billions of Kelvin. But at some point, the temperature reaches such extremes that the growing iron core itself begins to disintegrate.

While the core reaches its breaking point, the stellar surface endures its own turmoil. During the last hundred thousand years of a massive star's life (an unfathomable timescale for humans, but one that's short and sweet for a star), violent pulsations ripple through the star. Miniature explosions ignite like scattered bombs along the stellar surface, and intense winds strip the star's outermost layers. Sometimes, neighboring stars gorge themselves on the star, cannibalizing plumes of gas and plasma and ripping it apart.

When the core can no longer produce enough pressure to counteract gravity as iron elements disintegrate, the star buckles and collapses. Free electrons torn from iron elements get squeezed into other nuclei, forming neutrons and tiny, nearly massless elementary particles called neutrinos.

The core's outer layers fall inward until the subsequent pressure becomes so unbearable that the explosion rebounds, producing a shock wave and cascade of neutrinos that rip through the star and blow it apart.

This is a supernova.

The core, now composed of neutrons, is left behind in the explosion and becomes a highly dense *neutron star*. It floats in the massive star's tomb, encased in shells of gas that expand and fade over the course of centuries. Sometimes, the remnant core is so massive that it further collapses into a *black hole*, a pocket of space so dense that light itself cannot escape.

In a mere day, the brightness from the exploded star exceeds that of an entire galaxy. One exploded star can rival the brightness of 100 billion stars about the mass of our sun. Supernovae outshine even the brightest objects in our universe.

And they're responsible for the nature of the cosmos as we know it.

Elements produced deep within the bowels of supernovae float through the interstellar medium and shape the composition of the universe. Supernovae ashes are the seeds from which future generations of stars are born, for their elements get regenerated as the DNA of future stars.

We are the stuff of stars. The oxygen we breathe, the nitrogen in our DNA, the iron in our red blood cells, the calcium in our bones, and the carbon in our cells were once formed in the cores of stars.

Without supernovae, there would be no you. No me.

No dancing, no breathing. No words on this page, no leaves blowing in the wind.

No joy and no heartbreak, no thoughts and no love.

There would be no light.

Instead, because these stars die, the universe is alive.

I sit at the front of the physics auditorium in UT's Painter Hall in my freshman-year college physics class, my notes strewn around me and the professor pacing in front of the blackboard. When he speaks, it's with the unchallenged arrogance of someone used to getting his way.

He points to the equation he's just scribbled and waits for someone brave to volunteer the next step in the derivation. From above, the fluorescent lights glint off his forehead and shine down on my notes, casting the already indecipherable derivations in a blinding-white glare.

Nobody answers. He huffs, eyes narrowed, and scratches two Greek letters: a theta and a phi. These letters mean something to him, but to me they are simply shapes on a blackboard.

"What comes next?" He claps a meterstick on the theta.

I look around the auditorium to see rows and rows of twentysomethings, most of them boys, white, faces covered in puckered acne scars and patchy facial hair. Some of them are frat bros knocking out prerequisites, and others are the true physics nerds, the category that I fall into, the ones who genuinely want to understand our universe.

A flurry of hands shoots up around me, and I grimace, shutting my eyes.

To me, college physics seems impossible. We're nearly to the end of our first semester and I feel stupid all the time, slogging through my problem sets without the faintest idea of how to solve them. I try googling variations of the problems—this one is on an incline plane, that one dangling in the air, one is frictionless and one is floating in space—but the search results are less than helpful, just more equations and more derivations. My search history is now littered with parabolic flight paths and spring tensions—I hate it.

I've already declared myself as an astronomy major, but thanks to my dad's persuasion, tacked on the additional physics degree.

"It's highly attractive for the job market, and besides . . . you can't do astrophysics without the physics." He's right, of course, but I resent the decision all the same—I *hate* my physics class.

College physics is nothing like Mikan's observatory. Now my classes are in enormous, plain rooms with hundreds of silent students taught by stressed, overworked professors. The classes are not designed to pique students' curiosities but to weed them out. Our upcoming final is a three-hour multiple-choice exam, which makes me nauseated to think about. There's no room for creativity or mistakes, like there was in Mikan's class. Just perfection.

I'm realizing that physics requires an intuition that I don't have. If I just knew how to set up a problem or choose which parts contribute to the system I'm analyzing, I might be able to devise a strategy. But with most problems, I sit and stare at them for hours, stumped, slapping my pencil against the table in frustration.

From the corner of my eye, I spot one of the few women in our class raise her hand to answer the professor's question.

"The normal force is perpendicular to the contact surface," she says effortlessly. "So we need to take the cosine, not the sine, of the angle."

The professor nods, looking impressed for once, and erases the old equation, replacing it with a new one.

I tilt my head, mind whirring. Until now, I've felt too self-conscious to join the class study groups. I went once, at the beginning of the semester, and was, for the most part, ignored. Most of the attending students were men, and they confidently flocked together, bending their necks to peer at the problems and striding up to the whiteboard, one after the other, with self-assurance.

Office hours are difficult too. After our third class, I found my professor's office in RLM during the designated hour and knocked on his door with a list of questions. When the door cracked open and his watery eyes glimpsed my ponytail and armful of books, his head tilted to the right in surprise, as though I wasn't whom he expected to enter his door. At my first question, he huffed and ignored me; on the second, he simply stared at me until I quietly told him I'd look up the answer and escaped out of the office. I never went back.

But class is worse. Often, when he finishes his lecture early, he tells us about his failed marriage.

"What's the difference between a lawyer," he once said, referring to his divorce lawyer, while chuckling to himself at the front of class, "and a prostitute?"

Guffaws and laughter filled the room of men, and I looked down at my notebook, sick to my stomach.

"The prostitute," he finished, "tells you up front that you're about to get screwed."

I rolled my eyes, disgusted, and vowed to myself that I would only attend class if I really, really had to.

But if I studied with another woman, someone just as smart but less intimidating, maybe I'd feel more like I belonged. Maybe physics would make more sense. Maybe I wouldn't have to face this teacher alone.

For a split moment, I wonder if Grannell ever felt out of place like I do now when she was a computer programmer. Did she feel like an outsider as one of the only women in a sea of men too?

After class, as we file out of the classroom, I run to catch up to the woman who answered the professor's question.

"Hi!" I say as we exit the classroom and enter the atrium. "I'm Sarafina. Nice job answering that question today in class."

"Zoe," she says, sticking out her hand to shake mine. "Thanks so much. It's nice to meet you!"

We exit the building and move to the side of the staircase leading up to Painter Hall, watching students mill around.

"You're a physics major too, right?" I ask her.

"Yeah." She grimaces. "It's hard."

When I agree, she says more quietly, "Especially with this professor."

"No kidding." I shake my head, and a sense of relief shoots through me. She feels this way too.

Mustering up the courage, I ask, "Would you be interested in studying together? I could really use some help to survive this class."

"Oh my God." She heaves a heavy sigh and smiles. "I would love that."

In front of us, a group of girls wearing identical purple-and-white shirts, with DELTA GAMMA splashed across the front, see me and wave.

"Ready to go?" one of them calls to me. I call back to give me a minute and turn to Zoe.

"I'll text you to coordinate?" I ask, handing over my phone, and she types in her number.

"Definitely." She hands the phone back to me and smiles. "Can't wait."

The first week of college, I'd joined Delta Gamma, and although I had been reticent about Greek life, I knew I needed a group outside of my physics bubble. They're mostly nice, suburban white girls, smart and outgoing and eager to make friends, and although they'd been welcoming, I keep one foot out the door, anticipating the moment when I would no longer be welcome, no longer an interesting addition but a bizarre lack in judgment, a mistake.

In the desert of Texas whiteness, I am doubly self-conscious of my golden-brown skin, my curls that seem to be getting even curlier, my enormous eyes and sharp nose. *Sand Nigger, Bug Eyes,*

Dirty Jew, Terrorist. The old names echo in my mind every time I sit in the DG house for weekly chapter meetings, surrounded by my sisters.

Despite being a college freshman, I don't feel very *college-y.* I swap fraternity mixers and themed parties for nights in the library; I miss my sorority sister's birthday because I need to work through a problem set; I spend initiation coughing in bed with pneumonia.

Still, I'm grateful to be part of a group of women. Every day when I stride into physics class and look around at my classmates, my first thought is how different I am, how I don't belong. Although no sorority sisters are in my class, I see them around campus or in the library, and every time they wave and say hi. Because of them, I'm not alone.

And then there's Zoe.

Zoe has a quiet kindness, the type that sorority girls tend to overlook, and over the next several weeks, I learn that she loves cats, dreams of being a science teacher, is excellent at physics, and dislikes our physics professor as much as I do.

In the back of my mind, our upcoming physics final looms, but because of Zoe, it no longer feels impossible. We work together most nights, wading through problem sets and physics notes, swapping insights, and with her help, I develop strategies for solving physics problems.

I'm not dating anyone thanks to my workload, and my sorority sisters have stopped asking me to go out with them, but for the first time since I started college, maybe for the first time since I fell in love with the night sky, I feel like I'm not alone, as though the path I am trekking is suddenly filled with people—Zoe, Mikan, Dr. Wheeler, my dad—and together we are exploring the stars.

I think back to my dad's words from when I was in middle

school: "You can do whatever you set your mind to," he'd said, as if relaying a fundamental truth.

He's only half-right, I realize.

Without my support system, I think to myself, I would not be able to do this. They keep me afloat. They allow me to keep going.

On the morning of our final, physics formulae rise unbidden to my mind: $W = fdcos(theta)$, $v = vcos(theta)$ but only if it's in the x-direction, there's a sin if it's in the y-direction, and remember, the directions can flip depending on where the normal force points.

I meet Zoe outside of Painter Hall, and together we walk into the auditorium. Over the last several weeks, thanks to our study sessions, I slowly gained the physics intuition that I so sorely lacked. We learned to recognize specific types of problems and together developed a means for tackling each one. I still hate it, and as I sit down next to Zoe, the seeds of doubt rise unbidden in my mind: *Am I good enough to do this? Am I smart enough? Am I capable?*

Still, for better or worse I'm excited to be done with this class, to never have to sit through one of my professor's lectures again. Three of the other ten women already dropped the class, and one literally walked out of the room during lecture, muttering that she had better things to do than sit through sexist jokes.

I don't blame them—in fact, part of me is envious of their courage.

Our professor distributes the blue books and Scantrons methodically, handing the stacks to the students sitting at the ends of each row and allowing them to pass them down the aisles. Out of the corner of my eye I see Zoe reach across her desk in slow motion, and I make myself turn toward her to reach for an exam. She is smiling at me through her own fear and gripping my hand.

Her voice cuts through the crackling static of my mind. "We've studied for this. We've got this."

It's a gesture so tender that I become moored again and have to work to recenter myself in this seat in this auditorium on this campus, in the middle of Austin in a state called Texas, on a land-mass known as the United States anchored to the third planet from the sun, a blue-green marble floating in the sea of space on the outskirts of a galaxy called the Milky Way.

Days later, when I'm back at home for the holiday break, I'm sitting with my dad on our balcony sharing a celebratory beer.

"I know I got a B on the final," I am telling him. "But I honestly don't think I'm smart enough to do physics. What's the point of majoring in it if I'm miserable?"

From inside, the living-room lights switch off as my mom retires to the bedroom. I do a double-take, expecting to hear her footfalls as she walks upstairs, but force myself to remember that we are no longer in my childhood home. There are no staircases here.

Just before college started, my dad was laid off from JPMorgan, and my parents decided to move our lives—every boxed-up memory—into an eight-hundred-square-foot apartment.

I am still getting used to it. The pale, unmarked walls are entirely at odds with the sunshine-yellow kitchen of our old home. There, the staircase would have been lined with Christmas garland, the fireplace warming a living room housing a towering twelve-foot-tall Christmas tree.

Here, there is barely enough room for our sofa and a six-foot-tall tree.

"Did you know I was initially a physics major when I was in college at UT?" my dad says, and my eyes widen in surprise.

"Wait, really?"

He takes a sip of beer and chuckles. "Yeah. I got through about a year's worth of courses before I gave up."

"How have I never known this?" Frowning, my eyes cut to my dad, and he grins.

"You never asked," he says, winking.

The sky above darkens, and the stars begin to emerge. We're quiet for a moment, enjoying the night's chill.

"Believe me," my dad says, placing his empty beer bottle down on the table. "I know what it feels like to be totally lost. Physics is hard. For everyone."

"Well, I'm not enjoying it," I say, grimacing.

I don't tell him the unsaid part—that I don't know how much longer I can stand the sexism. I know that without Zoe, I never would have gotten through mechanics. Not just because she was nonjudgmental and kind, imparting some of her own physics intuition to me, but because I hated feeling so utterly uncomfortable in class. I don't want to experience this my entire career.

"I'm not surprised!" my dad agrees, oblivious to the internal dialogue running inside my mind. "These intro classes suck. That's kinda the point: they're designed to weed people out."

Then he turns to face me, an intense look in his eyes. "But you're not doing physics to become the world's best physicist," he says, "are you?"

"No," I admit, sighing. From somewhere deep in the forest just beyond our apartment comes the screech of an owl waking to the moonlight. I imagine that tonight it'll soar all the way up into the stars.

"Why are you doing it then? Other than to satisfy your old man?" I know he's leading me toward the answer that we both know, but still, I oblige him—this feels important.

"To study the stars," I mutter. "To become an astrophysicist."

I sigh, leaning back into the chair's cushion.

"What do you love about astronomy?" he asks me. I look up at the night sky, and a few stars are already glittering in the purpling dusk.

"I like feeling small," I begin, breathing in their starlight. "But it's not just that. I love asking questions about the universe—questions that people have asked since the dawn of time."

He nods, the crystal blue of his eyes glinting against the string of Christmas lights wound around the banister.

"There's something fundamentally human about that, isn't there?"

"Still, Dad, everyone understands physics better and faster than I do. *Everyone.*"

"Maybe right now," he says. "But I doubt any of them love the stars as much as you."

The next morning, my dad surprises me with the news that he's taking me to the McDonald Observatory. McDonald, UT's enormous observatory in West Texas, hosts public star parties alongside the research operations conducted on the biggest scopes. We'll be attending a star party together, he tells me, which means we'll be able to look through telescopes at some of the darkest skies in the country.

"I think the observatory will be a good reminder," my dad says while fiddling with his record player, "of why you're doing all of this."

The drive to the observatory is long. Eight hours of highway ultimately melts into long country roads, where ghost towns abound.

The closer we get to the observatory, the more alien a world this is: We pass a rancher on horseback who tips his cowboy hat in our direction before trotting to a cantina; abandoned trailers covered in mold and rust; the crumbling brick of a three-story, century-old jail; broken windows and the buckling roofs of deserted gas stations. In between forsaken shacks, one after the other in identical rows, sprouts native prairie dropseed in the broken cracks of the asphalt.

With time, the road tilts upward and we are once again in the mountains. And then I see them: the pearly domes of the observatory cresting into the sky.

I steal a glance at my dad, who is grinning and singing along to Talking Heads. When he belts the chorus, his eyes half close in bliss. I have the sudden overwhelming wish for the car to grind to a halt and for us to pause, as though frozen, to stop time before this moment slips away and slides into memory.

No part of me misses those long nights studying for physics—*this is living*, I tell myself.

A sign at the entrance greets us:

WELCOME TO THE MCDONALD OBSERVATORY
THE UNIVERSITY OF TEXAS AT AUSTIN, EST. 1933
VISITOR'S CENTER LOT: TURN LEFT.
ASTRONOMER'S LODGE: PROCEED STRAIGHT.

We pull into the parking area for the visitor's center just as the sun dips below the horizon. Behind the building, I spot four telescopes getting wheeled at forty-five-degree angles across the lawn in some predetermined arrangement. Two small domes stand on the edges of the lawn, their tops poking out just beyond the visitor's center.

It's already thirty degrees, and we expect it will drop below twenty by eight o'clock. I layer on ski gear: thick pants over my thermals, two sweatshirts, a parka, and a Patagonia shell, thick gloves, and finally my Delta Gamma ear-warmer headband. My dad slips his decade-old navy ski jacket on, and together we check in and walk to the lawn.

With each step, I can see more of the observatory sprawled beyond the visitor's center: two mountains, perhaps six or seven thousand feet tall, and three massive domes towering atop their peaks, two of them alabaster white, one of them silver, glittering as they reflect the last pale-pink rays of the setting sun. They dwarf the domes just behind the visitor's center, clearly reserved for research.

In the center of the lawn is a maze. Huge wooden benches are distributed in an oblong circle surrounding a slab of cement with speakers and a mic. A man breaks free from the crowd milling around and walks toward it, tugging his jacket tighter as he goes.

He grabs the mic and turns toward us, suggesting we claim one of the open benches.

"Welcome, everyone, to tonight's Star Party." His voice booms across the grass and echoes in the darkness. "I'm Frank, and I'll guide you through a tour of the constellations visible at this time of year. Afterwards, you'll get a chance to look through our telescopes, where you'll be able to see Saturn, Jupiter, and a few different galaxies and galaxy clusters."

My eyes go blurry and then refocus on the moonless sky, an uncountable number of stars beginning to peek through the haze of twilight. It looks as though there are clouds gathering in a distinct line cutting across the sky, and I wince just as Frank's green laser points to the same clouds.

"You might be wondering if we're going to be fighting clouds this evening, but I have good news: these aren't clouds." A few murmurs, and then he says, "That's the Milky Way galaxy. Our home."

I gasp, and feel the familiar lurch in my stomach, but this time it's due to awe, not anxiety, and I can detect the slight difference in my body: an airy drop of vertigo—this feels *good*.

It all hits me then, the utter vastness up above, the billions of stars and planets and galaxies—our very own galaxy! For an instant I am transported into the dark, and it is teeming, massive balls of roiling gas and plasma—stars—and there are dozens, thousands, billions of them, all lighting the universe, and then I am back on the wooden bench and there is the darkness once more, perforated by glimmers of starlight, glowing orbs hanging trillions of light-years away.

My dad takes my hand. I rest my head on his shoulder, breathing in his scent, all wood chips and pine needles and black pepper.

Frank's green laser traces constellations out of the jumble of stars. A great bear chases its cub from one hemisphere to the next, an eagle dives above a serpent bearer, a queen's crown rests atop her throne. Frank weaves magic into the sky, and it's as though I'm witnessing the scenes he describes play out: Hera transforming Zeus's child into the bear, Andromeda nailed to the sea-ridden stone.

Together, my dad and I watch, entranced.

As we make our way across the sky and our eyes adjust to the ever-deepening darkness, more and more stars appear, until the entire sky is aflame. It is unlike any sky I've seen.

The physics that I have done over the last few months is paltry compared to this expanse; but then again, so is the anxiety I allowed to consume me. I could not care less about the force diagrams I've nearly mastered, nor the derivations I've memorized, but as I look above, the realization anchors me that *this is why I do it*—this yet-unexplored, immense canvas of starlight.

I inhale a long breath and feel as though I'm returning home.

Frank says something. I feel my dad rising, and my neck snaps up, surprised.

"Dozed off?" he asks.

"Not quite."

He flashes a grin, a knowing look in his eye, and pulls me toward the nearest telescope. The only illumination comes from the red-lit domes on the far ends of the lawn. In the shadows, I see the outlines of people hunched over, desperately trying to warm themselves. The sight surprises me; I've forgotten about the cold.

Red light flickers. I walk up to a telescope and press into the eyepiece, straining against the urge to blink.

The telescope operator standing next to me asks, "You ever seen Saturn before?"

"No."

"Ah. Well"—he rotates a knob, and the eyepiece blurs out of focus—"take a look." My heart thumps once, twice, and then I see a perfectly ivory orb, encircled by oval rings and a half dozen specks of light—her moons.

It's utterly perfect.

"Holy shit," I accidentally mutter. My dad tsks, and the faceless telescope operator chuckles.

"Yep," he drawls. "That's her. See if you can spot the break in the rings; the Cassini Divide."

I squint until I spot the space where one of Saturn's moons has gravitationally pulled particles away from the planet's rings, forming a break that I am seeing with a telescope from a billion miles away.

I shove my eye up against the rubber, pressing so hard I blink back tears. I didn't expect it to be chalky white; it looks like one of the glow-in-the-dark stars my dad stuck to my bedroom ceiling when I was a kid.

Except: this is real.

After some time, the telescope operator clears his throat and I reluctantly break away from the telescope. He politely waves us away and turns toward the next guest.

The night is now so bright with starlight that we have no trouble finding the next telescope. This one is trained on Jupiter, and I gasp at its swirls and stripes, how the opaque orange looks just like it did in my childhood picture books.

My dad enjoys this almost as much as I do, delicately pressing his glasses into the eyepiece and sucking breath in through his teeth in awe.

"Remarkable that we can *see* it like this," he says when viewing Jupiter, then breaks free of the telescope and turns his gaze upward. We stroll toward the middle of the lawn, where Frank led the constellation tour.

"Thanks for taking me here," I say.

"You're welcome, little one." The red dome lights begin to click off. Car ignitions hum and their headlights twist away from the observatory. Something about the pinpricks of light above and that wood-chip-and-pine scent of my dad makes me think of my grandmother.

"How did Grannell do it?" I whisper into the night.

Caught off guard, he looks over at me. "What do you mean?"

"How did she manage being a woman in the sciences so long ago? Everything is already so hard, and I'm only just getting started."

Bemused, his lip quirks up.

"She did it because she loved it. She loved creating and solving puzzles. Remember your tangram?" He says this wistfully, looking back up at the stars.

I nod, looking up alongside him, picturing those wooden shapes as constellations.

"Of course."

"That was her way of sharing her love for mathematics with you. She started you early," he says, winking in the darkness.

"It wasn't only that though." He is growing more serious, the lines on his forehead deepening. "She wanted financial independence from your grandfather. She wanted to make sure she could take care of herself and her children—*no matter what.*"

Make good choices, Grannell's last piece of advice to me whispers in the breeze.

My dad runs his hand across his forehead, and the stars light his skin: he is glowing.

"I know how much you love astronomy, little one," he says. "You're the one who has the power to decide what you do with it."

CHAPTER VIII

BLACK HOLES AND THINGS TORN APART

Hiding in between the starlight of the cosmos are pockets of space so dark, so massive, that they threaten to devour the very fabric of the universe. Invisible yet violent, they bend spacetime itself, shredding objects that travel too close while swallowing passing rays of light. By warping spacetime into holes so dense that nothing—not even photons—can escape, they earn the title of most efficient information scramblers in the universe. Out of all the information necessary to describe the matter that falls into a black hole—chemical composition, temperature, size, and so on— only three pieces of information remain observable to our universe once that matter actually falls: mass, angular momentum, and electric charge.

Black holes litter the cosmos, so common that every single galaxy, on average, contains around 100 million. They're scattered amongst the stars, bending nearby light and feeding upon anything in their vicinity. Many wander the cosmic lanes between stars, roaming within their galaxies. And then there are black holes of the supermassive variety that reside in the centers of galaxies, engorging themselves on their surroundings and growing between millions to billions times the mass of our sun.

Black holes are the dark relics of stars unable to hold themselves together. When some massive stars reach the end of their life and collapse under the inexorable force of gravity, the gravitational force overwhelms even the pressure knitting together their core. The remaining mass—anything left of the star itself—implodes, collapsing into an infinitely dense region of space.

Because black holes warp the very fabric of the cosmos, they wreak havoc upon anything in their vicinity—even time itself. Objects that get too close, like stars and interstellar gas, fall toward the black hole, following the bend in spacetime like a bowling ball on a trampoline. As they near the black hole, they get caught in orbit and are slowly pulled apart bit by bit as the black hole feeds, devouring their innards and compressing what's left into a disk. Particles within the accretion disk ram into one another, heating up and emitting radiation that streams away into the cosmos.

Although black holes are invisible, their devastating effects are felt throughout the cosmos in the form of gravitational waves that ripple through spacetime at the speed of light. The energy released in the cataclysmic merger of two black holes is transported all the way across the universe, simultaneously squeezing and stretching spacetime, leaving an imprint of the black-hole merger forever embedded in the fabric of the universe.

And yet—despite the violence they exact upon the cosmos, black holes are fundamental to shaping the evolution of structure, like stars and galaxies, in the universe. Supermassive black holes at the cores of galaxies warp spacetime so much that the orbits of nearby stars are forever impacted. Accretion disks swirling around supermassive black holes release radiation that shapes how and where stars form. Blobs of material shooting out of black holes at nearly the speed of light form collimated jets of particles that seed the cosmos, molding the way stars form in distant parts of the universe.

Black holes are the disastrous ends of stars no longer able to sustain themselves. These are the invisible pockets of space that feed on a universe awash with light. And yet, even in their dark violence, the universe is made more beautiful.

I'm in my sophomore year of college, and I still hate my physics classes—even though I've resolved to stick with them to the bitter end. Physics is the hardest thing I've ever tried. Harder than Supernationals, harder than five-A.M. workouts, or my calculus BC class. It's even harder than that year of living by myself while in high school.

It's not just difficult. It's dry, stale. In physics classes, my thoughts aren't filled with stars; I am a robot, mechanically working through chapter after chapter of the assigned textbooks, their equations sliding like mud off the walls in my mind. Physics is the path to astronomy, I tell myself; still, though, I hate it.

But during the summer right after my freshman year, I took my dad's advice and spent three months interning at the McDonald Observatory, and under the star-strewn skies of West Texas.

Out there, while operating telescopes and watching the stars, night after night, I was reminded of why I do all of this. It all hit me at once—the utter vastness up above, the billions of stars and planets and galaxies, and how small we are in the grand expanse of the cosmos.

The universe felt big enough that I knew with certainty that my life, my very existence, could blink out and nothing would change; the universe would just continue on.

And the thought doesn't horrify me—it's comforting; liberating, even, a reminder that our transience, our atomicity, doesn't diminish how precious we are. It's simply that there's so much out there, timescales we can hardly fathom, and our existence is an exquisite blip. We are the single cell of an enormous giant, ten giants, ten trillion giants.

I begin the fall semester riding the high of the internship. I'm reinvigorated in my physics classes, knowing deep down that while I could not care less about the force diagrams or derivations, the starlit skies of West Texas—the unexplored, immense canvas of starlight—are why I am sticking with it.

One cloudless night in October, in a rare moment, I agree to go out with my pledge class. My physics classes this semester are going well so far, and I'm *happy*. I can afford to take a night off, I tell myself.

Six of us meet at my best friend Akira's apartment to pregame, and the room is coated in thick desire, that of young women wanting to be wanted, the chiffon dresses and silk tops swaying to the pop music someone blasts over the speaker. The white flash of cameras burns my eyes.

We sneak into the Blind Pig, their favorite bar on Sixth Street, colloquially dubbed Dirty Sixth. Used wristbands and red Solo cups litter the alleyways in between the rows of bars and night-clubs. The entire street is shut down for college students to swarm from one bar to the next, drunkenly swaying in the night's chill to the faint pulse of EDM.

Together, we down pickle shots and I fight against the urge to vomit, pinching my nose and grimacing. We cluster in a corner of the bar, and boys come up to us, one dragging Akira to the dance floor, another winking at Hannah and her winking back, Melissa following one to the counter to take another shot. The sickly sweet stench of sweat cloys at my nostrils.

Until now, I've had a couple of relationships but never one that's serious; they've ended with ghosting or cheating or simply because I'm too busy. The part of me that is desperately lonely is overshadowed by the other part of me, the one that works too hard and studies too much.

Singing lights the air with noise, ponytails swishing to the beat and beers sloshing down one another's fronts, and rather than join in I am so overwhelmed, so convinced of my own inadequacy, hyperaware that no boys came up to *me*, that here I am in my five-year-old jeans and humidity-spoiled curly hair and I am *no one*, nobody worth talking to, nobody worth *being*.

And then—from across the room, a man with dark-brown skin and even darker hair makes eye contact with me, eyes flashing against the green strobe, and prowls over.

"Hi," he says, stretching out his hand. "I'm Ahmad. Hannah told me that we should meet." A grin quirks the right side of his mouth, and my heart thuds unevenly against my chest.

Hannah, on the other side of the counter, glides over, a gleeful smile spread wide across her face.

"Yay! Finally!" She hugs both of us at the same time, pushing Ahmad and me awkwardly up against each other, and then steps back, admiring. He's tall, with a crooked nose and a smile that crinkles his entire face. I imagine that he's kind, with soft eyes like that.

"I've been wanting you two to meet for so long! Sarafina, Ahmad and I had a class together," she explains. A tall man with Greek letters inscribed on his chest beckons her, and Hannah winks his way, gesturing that she'll walk over to him soon.

"He's literally"—she slurs from the liquor, dragging out the *y*— "an angel. Ahmad the Angel, that's what we call him."

I raise my brow, and Ahmad flashes a winning smile.

"I hear you're Egyptian?" he asks, and Hannah leans over toward me, her ponytail tickling my collarbone.

"I might have told him a thing or two about you," she says loudly over the beat of the music, winking. "You two have sooo much in common."

Pink stains my cheeks, but my mind catches on his words—is he Arab American too?

"Half-Egyptian, yeah," I say, and Hannah hugs us both one more time before striding over to the beckoning man, now waiting with two lavender-colored shots and a sultry look.

"I'm Syrian. It's so nice to meet another Middle Easterner," he says, and I nod in disbelief. I can't believe my luck.

"Do you visit?" I ask him.

"I haven't been able to go back home in a long time," he says, and chews on the inside of his cheek. "I doubt I'll go back anytime soon given . . . everything."

A worn, faint memory surfaces, one of my mom and me in Cairo sailing on the Nile, our thirty-something Egyptian family members with us. Amr Diab plays on the speaker, and one of my cousins belly dances in the center of the platform on the bow. In the orange hues of sunset, beneath the infinite Egyptian sky, my mom laughs freely, singing along to the pounding of the drum.

She has not smiled in the same way since Egypt was plunged into revolution. Now when she moves through her days, she carries with her the weight of the sky. It is impossible to understand whether her mood, mercurial and heavy, is because she is separated from her family—and doesn't know when she'll be able to return—or because of something else. Our ancestors, she says, would not understand why we are so far away.

I'm speechless. Never in my life did I imagine I would meet a guy like this—like me!—at a bar, the child of Middle Eastern immigrants, who might understand what it is to be separated from their country. He has a knowing look in his eye, as though he can see inside the crease of my mind.

He pulls me outside, and in the quiet of the night he tells me about his dreams of going to graduate school, that he doesn't get along with his father, that his favorite thing to do is read. I tell him about astronomy and my complicated relationship with my mother. His eyes go wide, listening. When we stumble upon the fact that we both study at Caffé Medici and Epoch, two of my favorite coffee shops, I begin to laugh in disbelief.

In the distance, I vaguely register the bartender yelling for last call. We exchange numbers as I wave down a taxi, and before I slide into the backseat, Ahmad presses a kiss to my lips. The entire ride home, I feel like I'm flying, my night turned upside down and inside out.

Finally, I think, a guy who gets me.

Over the next two weeks, Ahmad and I spend most of our free time together. After classes, we hop from one coffee shop to the next, and I work on physics problem sets while he wades through organic chemistry.

He tells me that he wants to take a year off before graduate school to volunteer. He talks about what it will be like, taking care of people who will depend on him. At first I admire him, his compulsion to save others. It's only later that I realize that he also wants to save himself.

On the first of November, one month into dating, in between diagramming molecular structures at Caffé Medici, he tells me that he loves me, and I say it back. Later that night, I gather all of my clothes from my dorm room and drive them over to Ahmad's basement apartment, where I'm spending most of my nights anyway. Part of me registers that the relationship is moving quickly, but I tell myself that's just how it is when you're in love.

In the middle of November, when we've been dating for just over a month, Ahmad picks me up and drives us to Epoch for our nightly coffee-shop work session. It's eight o'clock and the sky is purpling, storm clouds gathering on the far horizon. Moonlight slants against gnarled old branches of cedars lining the parking lot, and fractures against the pavement when we pull in.

We nab the last table outside, and I plug in our laptops while Ahmad orders drinks: a black tea for him, an Americano with almond milk for me. Droplets the color of caramel stain the table when he sets them down.

Dr. Wheeler has invited me to join his research group, so in addition to my classes I'm now researching supernovae. I especially love our research meetings, when he pulls out papers on exploding stars and guides me through the important plots and figures. He says that asking questions is the job of a researcher, and not once do I feel stupid or small. The compassion he displayed during our very first meeting together, just hours after

Egypt was plunged into chaos, remains as pervasive as ever, and with each of our meetings the knowledge sinks deeper into me that Dr. Wheeler cares just as much for my well-being as he does for my career.

Still, my workload is piling up so high that I don't mind if most of my nights together with Ahmad are spent working. A small part of me prefers it, but I don't think too hard about why.

I slide out my laptop and pull up the code for the simulations I'm supposed to run, then stare blankly at the screen, annoyed. I don't even know how to download the code. My frustration mounting, I text another student in our research group, a senior named Steven, who responds rather unhelpfully: "Use wget in the Terminal."

I frown and type back that I have no idea what he's talking about.

"Search your Applications folder for the Terminal app, and then type wget —user-agent= "" mesa.com/sdk.tar.gz."

I do as he says, and five minutes later my computer whirs as MESA downloads and installs.

I grin and am in the middle of crafting a thank-you text back when Ahmad frowns at me. "Who are you talking to?"

"Just this guy in my research group." I barely look up from my phone, too focused on thanking Steven.

I'm met with silence, and I think he's dropped it and returned to his work until I press Send and tilt my head upward to meet his eyes.

He's still staring at me, eyes darkening. "Why are you two texting?"

Nonplussed, I say, "What do you mean? I had a question."

"Let me see your phone." His face a cold mask, he reaches across the table, hand dangling in front of mine.

"What? Why?" My heart begins to rattle against my chest.

"If you had nothing to hide, you'd give it to me." His hand

grabs mine, trying to wrestle the phone from me, and I feel his nails scratching into my skin.

"Ahmad, stop!" I hiss, ponytail whipping my face as I look around at the other customers, afraid to attract attention. A girl wearing a teal Kappa Delta T-shirt is staring at me, concerned. I look away.

"I swear to God, it's nothing," I whisper, looking back at Ahmad.

He drops my hand against the wood and pushes his chair back. The legs whine as they scratch against the floor tiles.

"If it was really nothing, you'd let me see your phone. Fuck you," he spits, "I'm leaving. Find your own ride home." He grabs his backpack and walks stiffly to the white Ford parked at the end of the lot.

What the fuck? I gasp for breath and shut my laptop, struggling to tug my backpack's zipper shut before following him.

"Wait, Ahmad! Stop!" I yell from across the parking lot. He doesn't answer.

"Stop!" I'm running now, kicking up pebbles of asphalt behind me. He gets into the driver's seat and immediately starts the ignition. Without pausing, he makes eye contact with me and floors the accelerator. The Ford jumps out onto Forty-Fifth Street, swinging a hard left before disappearing out of sight.

Heaving, tears sliding down my cheeks, I collapse onto the curb and stare blankly out into the street.

An hour later, thanks to a silent Uber ride across town, I'm standing in front of his door. My hands are shaking and I'm knocking once, twice, three times.

Finally, a silhouette looms at the window, behind the blinds, and the door cracks open.

"What?" he asks, still faceless, hiding behind the frame.

"I'm so sorry," I say, the words spilling out. "I should have given you the phone."

The door inches open, and finally I slip through. My heart is thudding. I try holding my arms to my chest to stop my body from shaking.

"Do you mean that?" he asks, eyes cold.

"Yes, yes, of course I do," I say, gulping for air, and now I'm crying in earnest. I'm not sure where the tears are coming from, whether they're a response to his anger or because deep down, I'm sure that this is all my fault. I should have given him my phone, let him read my text messages, and then he would see that I'm doing nothing wrong.

Or maybe I am.

Maybe he knows me better than I know myself.

He's insightful, and known to be kind—an *angel*—and for me to have elicited that sort of reaction . . . what does that say about me?

Ahmad reaches for the paper Epoch coffee cup and brings it toward us.

"How sorry are you?" he asks, black eyes not leaving my face.

"Extremely. I fucked up," I say. "I'm so sorry." I'm gripping my arms so tight that my hands are going numb.

He gives a slight nod and motions for me to come closer, and when I step toward his chest he drapes his left arm around my shoulder.

"Don't do it again," he whispers, embracing me, and then wraps his hand around my neck, pushing me up against the wall. His fingers press into the soft tissue—he is choking me.

In that moment, I feel the slick wetness of lukewarm coffee

spilling over my head and down my spine. Drop by drop, it slides onto my forehead and down my eyelids, and I blink back the murky brown liquid.

He's looking at me curiously, expectantly, waiting to see how I respond. Numb, I simply look back.

My mind, blessedly empty, is quiet, until the thought floats across it like a leaf blowing in the breeze, *I deserve this.*

One night of fighting, five, ten, twelve: I lose track. Anything can trigger him—the way I apologize, the text-message chain with the smart guy in my physics class, spending too much time with my sorority sisters, unexpectedly going to see my parents.

This is just how relationships are, I tell myself. *Ahmad the Angel*, I repeat in my mind when he punches a wall, fracturing plaster and drywall and puncturing his skin. His housemate says nothing, does not bang on the wall back.

We fall into an excruciating cycle. Night after night, we repeat versions of the same conversation until the early hours of the morning.

First comes the anger. "You bitch. You hurt me!" he screams at me in that basement apartment. "I'm going to hurt you."

Then, when I can't take his yelling anymore, when tears rack my body and I begin to shake uncontrollably, curling inward on myself and apologizing so many times that I go hoarse, he softens.

"I'm sorry, babe," he says, wiping away my tears. "I react with anger because I *love* you. For you, I'm expressing my emotions with such passion. If I didn't love you, Sarafina, I would walk away."

Sometimes, after his episodes, when he falls asleep, I slip outside and take my midnight walks. It's winter, but despite the frigid temperatures, walking calms me. Step, step, step: I fall into a kind of meditative state, my legs moving of their own accord across the streets of downtown Austin.

On my walks, I imagine that I'm back at the McDonald Observatory, looking up at stars that dangle through gaps in the mountains. Staring into the darkness, I replay his words in my mind—*You're so fucking stupid; you're a selfish piece of shit*—and feel bits of myself dislodging, floating away into the breeze, all the way to those mountains in West Texas and up to the sky above. If I hold my breath, I can hear the wind whipping across the canyons. Out there, it carries the whispers of starlight, alongside those forgotten, precious pieces of my soul.

Still, through the haze of sleep deprivation, I attend all of my classes, turn in all of my problem sets, complete simulation after simulation of supernova explosions for my research project. I am often so tired that my physics diagrams appear to quiver on the page, vibrating with the lines I slash across. I am never entirely awake, finding that it's easier to remain detached and bleary-eyed, drifting ghostlike through my days rather than feel too much.

It's unfortunate that physics and astronomy classes are mostly filled with men. Zoe's sequence of physics classes long ago diverged from mine, and now each man poses a threat to Ahmad. Despite my best efforts to reassure him, he is convinced that I will leave him for them. I learn not to exchange phone numbers for homework help. Study sessions and office hours are off-limits too. For the most part, I wade through my problem sets alone, with only textbooks and the internet for help. I am hating physics. My grades are staying afloat for now, but I enjoy none of it.

After an afternoon class in mid-January, I run into Akira, who pulls me aside. On the sidewalk, as students brush past, she tells me that she's concerned.

"I haven't been seeing much of you lately. Have you been sleeping?" she asks, noting my bloodshot eyes. "Are you okay?"

This isn't the first time a friend has posed this question to me since I've started dating Ahmad. Even Dr. Wheeler, during one of our research meetings, asked if I was feeling well. I wish I could tell them. I wish I could tell my dad, even my mom. But I know, without a shadow of a doubt, that he'll punish me if I tell anyone. This stays between us, he has ordered me.

"I'm fine," I say, offering a tired smile. "Classes are just kicking my ass."

Brow furrowed, she tries to ask another question, but I'm unable to continue the conversation—it hurts too much, keeping this from her. Instead, I make an excuse and set off on a run, all the way back to Ahmad's apartment.

One night in the middle of March, Ahmad and I are watching *Good Will Hunting* together in his apartment. We've already fought that day, when I'd told him my plan to apply to summer research internships.

"If you really loved me," he bit out, face disfigured by anger, "you wouldn't leave."

I've learned, over time, to simply surrender during our fights. If I dissociate, silently allowing him to rage until he runs out of steam, it doesn't hurt quite as much. Sometimes, while he screams,

I imagine I'm back at the observatory, lying on the grass and gazing up at the stars. Nothing can hurt me there.

"Okay," I whispered, shaking, when he slammed his fist onto his desk. "I'm sorry I mentioned it. I won't go."

After our fights, Ahmad is kind; generous, even. This time, he took me out to dinner at my favorite Mexican restaurant and, between dips of salsa, handed me a card in the shape of a bear that said in looping cursive, *"I'm so beary sorry—forgive me?"*

He seems so sincere when he apologizes that I feel as though I have no choice but to forgive him. Sometimes, he cries while he asks for forgiveness.

"I didn't mean any of it, I swear. This is just how my family is, my culture. You're the love of my life, we're *meant* to be together. Don't end this because I did something stupid."

His apology gifts—the dark-chocolate caramels, the oversize stuffed white dog, the piles of heart-shaped greeting cards and long, handwritten notes—are stacking up in the corner of our closet. I try not to look at them when I get dressed.

Even after tonight's apology, I can't focus on the movie. The overwhelming desire to travel, to advance my astronomy career and continue to learn, burns into me. Yet that future feels unattainable, his words sticking like burrs in my mind—*You're so fucking selfish, Sarafina.*

The louder his words echo, the faster my heart beats, until suddenly I feel it all the way through my chest and against the duvet. With every other breath, it beats erratically. My chest begins to ache, and now I'm gasping for air.

Mind spinning, sweat sliding across my skin, the thought sinks in: I think I'm having a heart attack.

I turn to Ahmad, who is next to me, looking on with concern.

"Ahmad," I bite out between gasps, "I need to go to the ER."

At the hospital, a nurse evaluates me immediately. She takes my vitals, notes my racing heart, and flags over a doctor to perform an electrocardiogram.

Ahmad is waiting in the wings of the room, watching as the doctor sticks electrodes all over my body. They're connected to lead wires hooked into a machine, recording the electrical impulses of my heart.

Dizzy now, on the verge of fainting, I close my eyes, willing the room to disappear.

Minutes pass, and I focus on breathing as evenly as I can, until the doctor gingerly removes the electrodes from my limbs.

"Good news," he says, gesturing for me to sit up. "Your heart is perfectly healthy."

Stunned, I stare at him. "I promise I wasn't making it up—" I begin, instinctively glancing at Ahmad, but the doctor cuts me off. He is watching me closely, registering my look toward Ahmad as though he can see right through me.

"I don't think you were." He smiles gently. "It sounds like you were having a panic attack."

The thought had never occurred to me. A panic attack could cause this? I hear Ahmad mutter "Thank God" under his breath.

"Panic attacks can manifest with physical symptoms—like a racing, uneven heartbeat." He surveys the room, noting Ahmad waiting in the wings, and moves slightly, blocking him from view.

Inside my mind, memories swirl with visions of doctors' offices and my drawer full of crinkled blue lollipop wrappers. I can almost taste that nasty, chalky drink that lit up my insides, the one they used to try to diagnose my phantom stomachache. I remember how sometimes I heave in breaths, unable to find enough oxygen. Were those panic attacks too?

Quietly, he says to me, "It might be helpful to see a therapist."

Blinking, I remain quiet. Although his brow is furrowed with concern, he finally says that I'm free to go.

When I'm discharged and we're silently driving back home, Ahmad wraps his hand around mine. In my chest, my heart begins to race once more.

I've chosen the weekend before my twenty-first birthday to tell my dad that I've decided against applying to summer internships. We are sitting outside of a restaurant under a canopy of trees, and I hope the celebration will temper his disappointment. Still, I hate to ruin this moment.

From across the table, I steal a glance at my dad, who is grinning. Seeing him so at ease, the setting sun flickering against his face, humming along to the music of the restaurant, makes my chest feel as though it's cracking open, rib by rib, breath by breath.

"But don't worry, Dr. Wheeler says I can work with him this summer," I say quickly. In my mind, this is the best-case scenario—I can stay home with Ahmad while continuing to do astronomy research. I ignore the nagging doubts that perhaps I should have just *tried* to apply. Trying would have meant betraying Ahmad, and I can't make him angry.

"I know it's not what you envisioned, but—" I start, but my dad cuts me off.

"I support whatever you want to do, little one. Whether it's becoming a waitress or the president; I don't care." His hand, resting on the skin of his forehead, begins to tremble. "As long as you're happy, I'm happy."

My heart skips a beat. I fight the urge to cry, something cotton-like welling in my chest until it feels like it will burst. I walk to the other side of the table and squeeze him, and he clamps his arms around me as tightly.

More than anything, I want to tell him about my relationship. But last time I got dinner with my dad, Ahmad picked me up and interrogated me in the car.

"Did you tell him about me? Did you talk about our relationship? Does he know we fight?"

"No, we only talked about you a little bit," I said, hoping to reassure him.

"I can see the lies on your face," he said furiously, gunning the accelerator. "How fucking dare you tell him about our private lives. They're *private.*"

So I stay silent and quickly wipe away my tears when my dad finally unlatches from our hug. He reaches behind his chair, rustling around in a bag I hadn't noticed, and pulls out a bottle of wine.

"We've been saving this since you were born," he says. "Happy twenty-first birthday."

Speechless, I tuck it against my chest.

"I love you, honey," he says to me.

When we say goodbye, I cry all the way home.

Sophomore year ends in May, and I stop eating. By July, I have lost twenty-five pounds. Food feels too greasy, too heavy, too much. My belly fills with the anxiety of knowing that inevitably, there will be another fight, another round of yelling and tears. Easier to just not eat, to instead become small and fade away.

I move back into my parents' apartment for the summer, and when I see my mom, her eyes rake over me with concern.

"Honey," she says, "you're not eating."

When I try to reassure her that I'm fine, she sees right through the flimsy lie and surprises me.

"I just need you to listen to me for a moment," she says, carefully choosing her words. "I don't think Ahmad is a good man."

Anger and shame flood my body, and I react out of instinct. "You don't know him!"

It is only later, when I'm curled onto Ahmad's bed after another one of his episodes, that I realize she was trying to protect me.

As the weeks blend into one another and July melts into August, our fights become more frequent and Ahmad's rage grows. Rarely a day passes without tears.

I begin to see my own therapist. Her name is Shina, and we meet weekly. Within weeks, she diagnoses me with an anxiety disorder, and I learn that it explains my panic attacks and constant worry, nausea, and nightmares. Ahmad, to his credit, looks worried when I tell him the extent of my anxiety.

"I don't want you to feel like this," he says to me one night. In answer, I nestle into him and ignore the pounding of my heart.

I'm careful what I share with Shina—nothing too specific about Ahmad, always making sure to tone down his anger, explaining that he only gets mad because he loves me. Sometimes, I have to pause our session to dive into my backpack and check that my phone is turned off, to make sure that he can't hear what I'm saying.

During one session, Shina gently suggests that if I want to stay with Ahmad, a couples counselor might help. For days, I muster up the courage to ask him to go.

After one of our fights—this time, I changed my email password and he screamed that I'm a lying bitch—he is in the apology phase of our cycle. When his eyes are finally clear, I test the waters.

"It might be good for us," I say, careful to emphasize *us*, "to learn strategies on how to communicate with each other. I hear couples therapy can be helpful."

To my surprise, Ahmad agrees.

"I know my anger makes you scared—even though I don't mean anything by it," he adds hastily. "So yeah, this could make sense."

My bottom lip trembles into a small smile, gratitude and relief coursing through me.

"Thanks, babe."

I call a dozen couples counselors, trying to juggle their availability with our lack of financial resources. Finally, I find one who accepts students on a sliding scale and can fit us in the following afternoon.

When I fall asleep that night, it is to the thought that maybe, just maybe, our relationship is finally going to get better.

The day we start couples counseling is unbearably hot. We take two separate cars to our therapist's quaint white house nestled into the hills of Westlake—I'm visiting my parents across town, and Ahmad is coming from campus.

At first, the session seems to be going well. Ahmad is describing how difficult it is to trust me, how he is terrified that I'm going to leave him for someone else. I stroke his knee as he talks, gently reassuring him that I'm still here.

When the therapist asks me directly how I feel in the relationship, I say how much I love Ahmad.

"But to be honest," I say, "his anger can be a lot."

I'm careful not to say too much, but Ahmad still shoots me a burning look—I know that he will punish me for this later.

"Perhaps," she says to him, "there's another way to express your fears."

He tries to justify himself.

"My words might feel harsh," he says, "but I'm trilingual, and this is just what we say in my culture. It might translate differently in English and just sound worse than they are."

From across the room, the therapist narrows her eyes and jots something down in her notebook but says nothing. I say nothing too.

When the session is over and we reach our cars, finally out of earshot of the therapist, Ahmad hisses to meet him back at his apartment.

"She's incompetent," he says furiously, slamming his door into mine. He watches as aluminum hits aluminum and slams it again, chipping the blue paint on my car. I stand silently, tears gathering, until he slides into his front seat and accelerates away.

Slowly, I begin to drive toward his apartment. I take back roads, desperately trying to delay the inevitable fight. But not fifteen minutes into the drive, he calls. When I answer, his screaming echoes throughout my car.

"Where the fuck are you? I'm waiting—get over here now!" He must have just made it back home; in the background, I can hear the apartment door slamming shut.

"You talked to her alone, didn't you?" he yells, referring to the therapist. "You got her onto your side."

"No, I swear—" I try to say, but he is hearing none of it.

"You're a fucking LIAR! I'm going to kill you! You're *destroying* me." He's screaming so loud that I thumb the volume down, yet I can still hear him as though he were beside me in the car.

"I'm going to fucking kill you, do you hear me? I'm going to kill you, and then kill your parents. I'm going to murder you and your stupid fucking family."

Tears burst forth and I'm hysterically crying, sobbing for him to stop, please stop.

"I will do anything," I am yelling back at him, needing him to hear me. "Please, just stop."

"Where the fuck are you? I'm here waiting." In the background, I hear a slamming noise—he is throwing something against a wall.

"You hear that?" he screams. "That's your dad's stupid bottle of wine, and you know what I'm going to do? Fucking break it."

I can barely see, I'm crying so hard. The car swerves, and my body is responding entirely of its own accord, launching forward and pressing the gas pedal.

"I'm going to kill you." He is screaming and repeating it over and over. Still, I drive toward him, until I hear him say, in a level, horrified voice:

"Fuck. Someone called the police."

And the line goes dead.

When I arrive, three police cars are parked in front of his apartment building, flashing the white stucco in red and blue lights. I stay in my car, numb, watching. Two officers crowd his door, one taking notes and the other speaking into a radio. I can hear Ahmad's voice calmly speaking with a third officer out of view. Heaving, I slide down my seat, hiding from the police—from him.

I don't want the police to find me, to question me. I don't know what I'm allowed to say, don't want to put him in any danger or anger Ahmad even further.

Half an hour later, the lights flash off and the police cars pull away. Mercifully, they never spotted me. When I get out of the car, my feet take me across the pavement and down the stairs until I blink and am in front of his door.

"I'm here," I say, and the door cracks open.

He's a mess, tears streaking his face and clothes in disarray. In the kitchen is the shattered bottle of wine, the red liquid already staining the white of the countertop. I look upon it all, numb.

"One of my neighbors called the police," he whispers, voice breaking. I follow him to the bedroom, where he collapses on the floor and curls into the fetal position.

"Apparently they heard me saying—" He does not finish the sentence, but I know what he is leaving unsaid: they heard his death threats. The sliver of me that is not yet numb wonders if the neighbors have overheard other fights or spotted the bruises discoloring my clavicles. That part of me hopes that they have.

"I can't believe it. My future is ruined," he's crying. "This is going to ruin my whole fucking life."

Something vicious twists inside me, a deep-seated anger curling in disgust. In this moment, after death threats and police, he's thinking about his future?

Still, I kneel down to rub his back.

"It's going to be okay," I whisper.

Over the next three days, I am sick with fear that Ahmad will kill me. At night, when I'm back at my parents' apartment, I slide a chest of drawers across the floor to barricade my door. Every few hours, I wake in feverish terror that he is here.

Still, I tell no one about the incident except for Shina. For the first time since we've been working together, her composure breaks.

"I'm worried about you," she says.

Four days after the incident with the police, she emails me.

Sarafina,

I am concerned about you and I care for you and your safety.

If you ever need me or need to talk, please feel free to call me.

If you ever feel you are in immediate danger, call 911.

I'm sitting alone outside of Whole Foods when I receive the email, and my heart drops into my stomach. Over and over again, I read her words, trying to wrap my mind around what she is saying.

For her to put this into writing, a small, terrified part of me realizes, she must be genuinely concerned for my well-being.

And another part of me—the one that has been normalizing Ahmad's behavior all this time, registers this as evidence; proof that his behavior is not okay.

Perhaps, that small voice in me is saying, *you don't deserve this.*

Fear drives me to meet Shina for an emergency therapy session. Together, we form an escape plan. Hands shaking, I write down words of how to break up with him and where to go after I leave.

Immediately after the session, I meet Ahmad in a parking garage on the edge of campus. I am moving outside of my body, as though watching this happen to someone else. Waves of exhaustion and ripples of fear are crashing into me, but no sadness. None.

He knows something is wrong, sees it in my face. When I tell him I am breaking up with him, he begins to yell, just as Shina predicted.

Following her directions, I cut the conversation short and start the ignition. He palms at my door, and I look at him from behind the window.

"Please," he begs, "please don't leave me."

Some previously unknown reservoir of strength deep inside me pushes me to ignore him, to press my foot to the pedal and, despite his pleas now turning into screams, drive away.

The night of leaving Ahmad, I go to my parents' apartment and confess the abuse. When my mom cries, I cry too. Something about this abuse is deeply familiar to her, and although I don't entirely understand why, it's almost as though she is reliving something in her own past in this moment too. Somewhere within me, I am proud of both of us for getting out.

Within minutes, my dad, pale and stone-faced, reconfigures my email to block Ahmad's address and filter his messages to the trash. The next day, he takes my phone to the store and changes my number.

I spend the next few months learning how to heal.

I reintroduce normal meals into my routine. I've lost thirty pounds since I started dating Ahmad. Still, physically, I heal quickly. I rebuild muscle and regain fat. I grow stronger and run every day.

I stay at my parents' apartment, too afraid to live alone. Sometimes, when I wake in night sweats and I gasp for fresh air, my mom is awake too. She brings me a glass of cold water and sits with me when I cry.

My dad and I begin to stargaze again. We sit outside on our balcony in a companionable silence, savoring the feeling of being together.

One night in October, when the air is chilly and we're watching the stars, my dad turns to me and says seriously, "I'm so sorry you went through that . . . abuse, little one."

I reach over for his hand, and for a split second I'm five years old again, alone and anxious, and my dad pulls me into our backyard.

Would you like to stargaze with me?

I see myself sliding open the glass door and following him outside, little legs struggling to keep up with his. He pulls out the binoculars and sticks them to his glasses. I sit beside him just as I am now, eyes trained on the sky. When I inhale, I breathe in the faint, comforting scent of pine needles and woodchips. Everything is quiet.

The understanding settles into me that, no matter what, I need this; I need the night sky just as I need food and water and *my dad.*

With them, I am safe.

With each breath I'm struck anew by the knowledge that this is home: this pine scent, the stars stretching into the darkness. The night enveloping us as the Earth turns on its axis until dawn rises once more.

$$L \leq L_0$$

$$u(v,T) = v^3 f(v/T)$$

$$u(v,T) = c_1 v^3 \exp(-c_2 v/T)$$

PART III

FATES

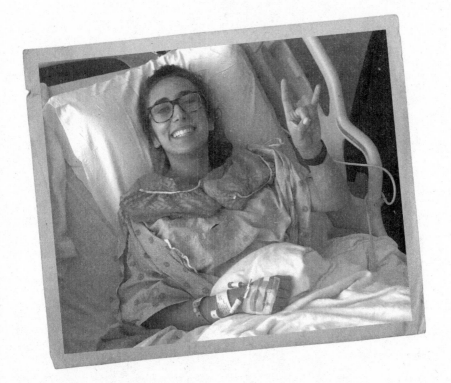

ion trail

gas

nucleus

dust tr

comet

$$u(v,T) = P_v U = \frac{8\pi h v^3}{c^3 (e^{hv/kT}-1)}$$

CHAPTER IX

COMPOSITION OF
THE UNIVERSE

For as long as humans have roamed the Earth, we have sought to find our place in the cosmos. From the city-states of ancient Greece to the soaring capstones of the Egyptian pyramids, across the deserts and towering mountains of ancient China down to the rolling plains of Mesoamerica, humans have sought to understand how the universe works. They developed mathematics to trace the motions of the planets, estimated the circumference of the Earth by walking from city to city, created star tables and timekeeping codices, and even recorded celestial events like Halley's Comet, supernovae, and eclipses.

With time, we have refined our models of the universe. Using ellipses, Johannes Kepler reconfigured celestial motions. Galileo revolutionized Copernicus's heliocentric model of the solar system by discovering that the sun, not the Earth, is the body around which all other elements of the solar system orbit. Isaac Newton developed the theory of gravity, which was later supplanted by Albert Einstein's theory of general relativity.

Discovery by discovery, we paint in the gaps of the picture of our universe; and yet somehow, with each brushstroke, that image morphs, evolving into something ever-changing, new, and unrecognizable. The universe that Kepler and Galileo, Copernicus and Kepler, Newton and Galileo, and even Einstein understood is different from the one we know today.

Today's understanding of the universe is unsettling. It is not one that fits in a tidy little box with neat lines and a perfect lid. Our universe is mystifying, complex. It defies expectations.

For starters, our universe is not a static, enclosed entity. Our universe is *expanding*. From everywhere all at once, the fabric of spacetime is stretching away from everywhere else like an inflating balloon, carrying galaxies along with it. Photons traveling the lanes of the cosmos are stretched along with spacetime, their wavelengths growing ever-longer, or redder, thus *red-shifting* with the expansion of space.

Our universe isn't expanding *into* anything. To our knowledge, there is no extra dimension around the universe; rather, space itself is expanding, causing the space between galaxy clusters—the largest gravitationally bound objects in the universe—to get bigger and bigger with time.

And this leads us to the following unsettling conclusion: there's no center to our universe. Everywhere is the "center" because everything everywhere is moving away from everything else, all at once.

But the universe isn't just expanding.

It's accelerating.

With each passing moment, an unknown repulsive, persistent force dubbed "dark energy" is stretching the fabric of the universe. Dark energy is a fundamental property of space itself; invisible, smooth, and constant—and yet we are entirely unsure what it truly is.

And then there is dark matter—the invisible, clumpy matter that binds galaxies together. In many ways, dark matter is the corollary to dark energy: where dark energy stretches space apart, dark matter knits matter together. They are both invisible—neither interacts with radiation or light—and yet they are ever-present, dark matter acting as the cosmic glue for large-scale structure formation, and dark energy a principal ingredient in the universe's evolution.

The afterglow of the Big Bang, known as the cosmic microwave background, is imprinted on the fabric of spacetime, a relic of radiation from when the universe was extraordinarily hot, dense, and smooth. By mapping its bumps and irregularities and comparing with galaxy surveys, scientists have found that 70 percent of the universe is made up of dark energy. Meanwhile, 25 percent of the universe is dark matter.

Just 5 percent of the universe is ordinary matter.

That's the ordinary matter of everyday life: your hair and clothes, your atoms and organs, the food you eat and the dogs that kiss you, the air and the sea, the sun and the moon. Everything we know—everything we see—is just 5 percent of everything in the universe.

The remaining 95 percent of the universe is stuff that we can't see, don't yet understand. An extraordinarily vast portion of the cosmos is still unknown. Despite the technological advancements of the last century, even with computers at our fingertips and the worldwide internet and space-based observatories mapping the far reaches of our universe, there is still so much that we don't understand.

We have grown leaps and bounds since the days of the ancient

Greeks and Egyptians, even since Copernicus and Kepler. But in many ways, we are still novices playing with toy models seeking to understand the stars.

At the end of the day, we are on a lone planet dangling in space, orbiting our sun amidst millions of other stars in a small corner of a galaxy in an ever-expanding universe.

It is in our nature as humans to seek meaning in the stars.

But sometimes, the answers aren't the ones we were searching for.

Above us, the sky explodes.

In the roiling blackness, red and blue lights pop in and out of existence: *bang, bang, bang.* A pillar of smoke twists into the sky, tracing constellations that disappear into the wind.

It's the beginning of December and I'm in my senior year of college, finishing up my dual degrees in physics and astronomy.

My dad and I are perched on the edge of a cliff on Highway 360. It crests just over the Colorado River, looking out over the sparkling Austin skyline, rolling Hill Country, and, most important, the stars.

Since my breakup with Ahmad, it's become our favorite place to see each other.

Over the last two years, I've stopped frequenting Epoch and Caffé Medici, opting instead to study at home behind the safety of a locked door. On campus, I spend my walks between classes looking over my shoulder, anxious that Ahmad will spot me. I briefly consider filing a restraining order but am too afraid of his retaliation to proceed.

While the physical abuse has melted into the haze of memory, emotional healing is the hard part. I go to therapy twice a week, and when we talk through Ahmad's abuse, Shina diagnoses me with PTSD. My body still remembers what it is to be pushed and shoved, screamed at and belittled, manipulated and afraid.

I tell her that my inner self feels shredded apart. No longer do I trust myself.

It will be a long, long time, I think, before I am ready to date again. Before I am able to feel anything beyond the emotional tremors from the abuse.

On the overlook, an icy breeze gusts, speckling our faces with the spray of a surf boat roaring far below. I lean into my dad, inhaling the comforting scent of pine and smoke. After a moment, he gently pushes me away. Through the mask of night, I spot his lips turn upward in a grimace as he readjusts his position on the stone.

"We should go," he says.

"It's only ten."

"I don't feel up for staying much longer." Groaning, he swings his legs from over the cliff and towers up tall. I hear a knee pop as he straightens himself out.

"But we never leave this early." If it had been another night, a normal night, we would only now be settling into our spot in the rocks; these nights are sacred.

He moves as though he can't hear me, lumbering slowly back to the trail. The sound of scattering rocks clangs like alarm bells.

I spring up and look down once more. Pools of darkness quiver in the river. Shimmering red and green lights reflect off the firework-smoke clouds gathering on the horizon. Attendees of the Christmas festival on the opposite shore spray champagne into the sky. It seems impossible to me that they don't feel the fear slipping in too.

My dad leads the way down the trail. My ankle twists on a

knobby root as I hurry to catch up to him. Noise from the crowd is subsumed by the thickening smoke, until we are plunged into an eerie quiet. I feel as though we are underwater, the air thick and heavy, the darkness suffocating.

When we finally reach the car, my dad eases his body into the driver's seat. The black leather is cold, and the worn cracks dig into my jeans.

He turns the key and the car rumbles to life. Shadows follow us as we merge onto the highway. He's clutching the wheel so hard his knuckles go white.

"Are you feeling okay?"

He sucks in a breath. "Not really."

I look at him, really look at him. His eyes are scattered, lips cracked. He shifts his position once more, now leaning forward against the wheel.

"I have some . . . pain," he admits. "At first it was just when I sit down. Now it's whenever I move. Whenever I breathe."

Cars pass, their headlights blazing against the side mirror. I shut my eyes.

"How long has this been going on?"

"I didn't want to worry you," he says. "About six months." My eyes fly open. Passing streetlights flicker. Their gleam blends into one long blurry line that stains the sky.

A carousel of memories plays: Dad visiting me at my internship at Harvard last summer, regularly sitting down on walks across campus. Finding wooden park benches to sit on tucked between churches along Mass Ave—not just once but three times. The Tylenol he popped nightly for the phantom pain in his lower back that he insisted wasn't a big deal. The Cape Cod hotel bed with frayed sheets. He stayed tucked between them for the entire weekend.

Hands shaking, I splay my fingers along the leather door trim and feel it cool my skin.

I'd asked him about it at the time, but he'd simply blown me off. "It's fine, honey. Don't worry about me. How's your research going? How are classes? Any better?"

I'd pushed him that weekend in Cape Cod, when I'd returned from the beach to find him groaning in the fetal position on top of the sheets. I remember how they rolled around his limbs, sweaty and limp. The sour smell that hung in the air.

"Nothing's the matter. I promise, little one," he'd said.

The car swings right, and we're entering our apartment complex. From the asphalt pathways dart shadows that coil into monsters. They follow us in the darkness, all the way until we enter the garage, when the fluorescent floodlights shatter them into a thousand pieces.

"They think it's a cyst," he says. "I'm having a procedure to remove and test it in a couple of days."

A cyst can be benign, I remind myself. Yet despite the logic, I feel the terror ticking through me.

We shut the car doors and walk down the sidewalk, up two flights of stairs, and wait in front of our apartment door while he rummages through his pockets. His hand, as he fumbles for the key, is cracked and white, and his fingers are shaking. He drops the key once, twice, but then he has it and we are inside.

When I crawl into bed, I remind myself to be logical. Remember, once again, a cyst can be benign.

My eyes flutter shut, and the demons following us through the shadows find the cracks in my eyelids. I squeeze my pillow and try to ignore the team of horses galloping across my chest.

I promise, little one.

Three days later, as the sun crests over the horizon and rays of light perforate swollen clouds, we drive to the hospital and my dad has surgery on his prostate.

It's a quick procedure, and while we wait, my mom and I walk laps in silence around the parking lot. The sky eventually cracks open and rain pelts our curls, but we continue walking, around and around and around, under the blue ER awning, past three rows of ambulances and beyond an empty police car, until the sidewalk runs out and we turn around and return to the sliding glass doors.

At 5:04 my mom's phone rings.

"This is Dr. El-Badry," she stammers. A muffled voice responds, and she puts a hand against my chest to halt our walk.

Moments later she smiles, hangs up, and says, "It went well. He's out, let's go see him." She lets out a long exhale and I trail her through the sliding doors, lightheaded with relief.

We find him on the fourth floor, in a private room the shade of sand. Two nurses are already there, one helping him sip water from a straw and the other adjusting his pillow. Long, clear tubes sprout from his waist, filling with a dark-red mixture of blood and pus. His eyes are foggy and he fumbles for his glasses, and then he squirms, moans, and the nurse says something I can't understand, and suddenly he's bellowing in pain.

The sound cracks through the air like thunder, and I feel it all the way through my chest and in my bones.

I have never heard him make a sound like this.

I reach for my mom's hand at the same moment she reaches for mine, and we're squeezing each other as one of the nurses presses a blue button on the side of the hospital bed. A bespectacled doctor rushes in. The other nurse turns toward us and gives us a sad smile and asks us ever so courteously, could we please wait in the hallway while they straighten this out?

Hearts thumping, we walk out, and the door slams shut behind us. My mom heaves a sigh and leans against a windowpane; she looks as though the past three days have aged her ten years.

I cannot think. A buzzing fills my ears and everything is far too bright and smells of bleach and blood. My stomach turns.

Nurses and doctors move around us, scribbling on yellow notepads with crimped, illegible cursive and carrying dozens of pill bottles, ushering one patient on a stretcher and another in a wheelchair, down the corridor. A woman yells in another room, and my nostrils flare at the overwhelmingly bitter scent of bleach and I swear I am going to faint.

The door cracks open and the nurse gives us that overly courteous smile once more and tells us that they've doubled his morphine and he's resting now. My mom responds and they have an entire conversation and I hear none of it.

His screams of pain will haunt me for the rest of my life.

Hours later, he wakes up.

The nurse bustles around the IV, fixing tubes and shifting his catheter placement ever so slightly. She tells us they've sorted out his pain-medication dosage. His eyes are glassy now, pupils dilated and unfocused, but he's giving us a lopsided smile.

My mom collapses in the ugly tan faux-leather chair next to his bed, and I stand awkwardly at his feet, waiting for him to speak.

"Honey! How's studying?" He's slurring his words. The irony of the moment seizes me: despite the liters of blood draining from his body and into half a dozen tubes; despite the immense pain he must be in and the anesthesia still clogging his veins;

despite the raw fear he must feel—his first thought is of me and my studies.

"It's fine, Dad. It doesn't matter. How are you feeling?"

"I'm fine, just fine. Tell me, have you finished up your finals yet?"

"Yep, just getting ready for the astronomy conference."

"Did you submit all your grad-school applications?"

"Yeah, Dad, you know this," I say. "Remember?"

The nurse brings in a tray with a glass of water, grape Jell-O, and a half serving of rice. My mom looks at it and blanches; I do too. My dad doesn't seem to mind, he pokes the Jell-O with his fork and grins as it wiggles.

"Good, good. Hey honey, guess what?"

"What?"

"I told the nurses a joke—" he starts.

"Oh God." I roll my eyes and grin.

"You know, that one about Peter Pan?" He smirks, glassy eyes crinkling with laughter, and my chest feels as though it's breaking open: it is so good to see him like this.

"Why is he always flying?" He grins boyishly. "He *neverlands.*"

"And get it, it's extra funny because I'm Petey! Peter Petey!"

Suddenly, he looks decades younger. The tubes and monosyllabic beeping of the IV machine fade into the background: he could be sprawled on the couch, sipping a cold Sierra Nevada and watching *The Colbert Report.*

A giggle bubbles out of his chest and he grunts, *"Ow."*

My mom springs up from the chair and her hands go to his tubes, his chest. "Peter, stop moving!"

Waves of dizziness break over me. I slide into the abandoned chair and grip the armrests like a life raft. *Beep beep beepbeepbeep*, the machine tings. A nurse bustles in and politely shoves my mom out of the way.

My dad grimaces and the light in his eyes is gone; he is a hospital patient again. We wait until he falls asleep, the morphine overtaking him, and I kiss his cheek. His skin is paper-thin beneath my lips.

We make our way back downstairs, through the sliding glass doors, past the ambulances and into the Volkswagen. Tiny flecks of water cascade from the darkness, a rainstorm peppering our windows.

We drive most of the way home in silence, until we're on Pennybacker Bridge and my mom turns to me. "Do you remember what you used to say when it rained?"

I shake my head.

"You said"—she sniffs—"that God was crying."

A day later, we pick up my dad from the hospital. He wobbles into the car, zonked on painkillers, gingerly avoiding the tubes stitched to his skin and the catheter poking out of him. He falls asleep almost immediately when he settles into the passenger seat, and once we get home he retreats straight to the bedroom.

At six o'clock, my mom's phone rings.

Primly, she strides to the bedroom with a yellow legal pad in one hand and phone in the other and slams the door shut behind her. The baby hairs on the back of my neck prickle, electrified.

Desperate for distraction, I open my laptop and pull up my poster for the astronomy conference.

Based on our simulations, massive stars appear to retain trace amounts of hydrogen prior to explosion, I type. The percentages, definite and unambiguous, calm me: 4 percent hydrogen in one simulation, 10 percent, 8 percent. The black-and-green computer terminal blinks in the background as long strings of numbers roll across the screen.

I've learned that astrophysics is meditative. Derivations force my mind to bend in unfamiliar ways, cranking gears and bending axles; step by step its wheels tick. Somehow over the last three years of impossible courses and infinite problem sets, I've emerged from a thick, low-lying fog and grown my physics intuition.

It's imperfect, often winding me right back to where I started, but it's something. I'm learning to sharpen my instincts, to make them my own.

But: it's there. A jolt of realization, scribbling numbers and theorems, tearing through textbooks, and I have solved one problem, two problems, three!

It is during my research when I make the most progress, when I'm running simulations and seeing with my own eyes what tweaking parameters does: increasing the Dutch wind by 0.5 hastens the stellar explosion by a factor of two; decreasing the mass plunges the simulation into havoc.

A calmness settles over me when I hunch over my laptop and create my simulations. Like the fantasy novels I adore, research provides a portal to another world, and in this world, rules are still being written.

Soon enough, I tune out whatever's going on behind that bedroom door. Like Schrödinger's cat, that phone call exists in a state of unknowing, dangling in the in-between, until the door swings open and the spell is broken.

My mom is standing in front of my computer, eyes strained.

"Please come to the bedroom—your father and I would like to talk to you." I nod and beads of cold sweat gather on my brow. I raise a shaking hand to wipe them away.

I shuffle behind her, past our hanging list of hieroglyphics, past the family photo of us in front of the pyramids, past my life as I know it.

My dad is propped up on a pillow, the bulbs attached to the end of his tubes dangling off the side of the bed. Terror thrums through my body. The hushed phone call. Tears already falling from his red-rimmed, swollen eyes.

"What's wrong?"

"That was the doctor," he says. "He reviewed the biopsy and scans, and, honey . . . I have metastatic prostate cancer. Stage four."

All at once, the air is sucked from the room.

Cancer. Stage four. Terminal.

The words float through me, ghostlike. I feel nothing: no fear, no surprise, no devastation, just an icy emptiness flooding my veins.

One of his drains is clogged, the mucuos mixture of pus and blood clotting the entrance to the bulb. My mom sidles over to the bedside table and gently runs her hands along the rubber, squeezing the mixture until it pops out of the tube and into the bulb.

My gut heaves, and I taste vomit.

She perches on the end of the bed and begins to cry, delicately at first, then transitioning to violent, wracking sobs that shake her entire body, throwing her curls askew. The kohl lining her eyes runs like rivulets down her cheeks.

My dad simply lies there, silent, fragile, frozen to his pillow. Despite his girth he looks skeletal, the skin of his neck sagging into his pillow. How can a man with such vigor be brought to *this* so abruptly?

"Stage four," I repeat. "What does that mean?" He gives a brief nod of recognition, understanding that analytical questions are easier than emotional ones, and I swear I see a flicker of appreciation in his eyes.

"It means that I will probably die of cancer," he says, voice trembling at the end.

"But," he continues bravely, "it doesn't mean I'm dying tomorrow. I'm simply more likely to die of cancer than, say, getting hit by a bus. The average prognosis is two years, but who knows, anything can happen. Could be five, could be ten!"

It is too much. My head is swimming: how can our entire lives change in one split second? Impossible that just days ago we were stargazing on a cliff, and now I have tumbled over the edge and fallen into the choppy waves cresting below.

From the bed rises a translucent ghost of my dad, the one who existed up until this moment, and his face is transfixed into a heartbreaking smile as he floats out of the room and into the black void of night. The figure remaining on the bed is born of this moment, a new Peter Kevan Nance who has stage-four cancer and is poised to endure a cocktail of toxins and shower himself in radiation for the slim chance of hanging on to this new life for much longer.

My heart rattles in its cage; I am beginning to hyperventilate. I try to remember how to breathe: inhale—exhale—inhale—exhale. My mom begins to wail in long strings of Arabic, and my blood is thrumming so hard that I feel it in my ears. I whirl to face her.

"Can you just let him talk?" I shout, burning at the indulgent emotion, how it is *his* decision to cry or scream, not ours, and we should follow his lead like the dutiful warriors we are.

Caught off guard, she hiccups a cough. My dad, too weak to forge peace between us, takes my hand in his and gently rolls his index finger over mine.

"Let me just go into what the urologist said," he says. "He's referring me to an oncologist in Austin—though he suggested I look into MD Anderson. The oncologist will come up with a plan of action, but he suspects that I'll need to begin chemo immediately."

As he continues saying words that I can no longer process, I hover somewhere above myself, looking down on the events un-folding below in horror, desperately floating upward toward the stars to escape this moment, whatever this moment *is*, for I want none of it. Up there in the cosmos is beauty, up there makes sense, up there are stories to be discovered and ancient histories of the universe that make this moment small, so small. There are stars dying *this very moment,* and other stars just being born; there are suns erupting and planets breaking and chaos, oh so much chaos, and this is just like that, this is, I think to myself, just another moment of change, a tiny blip in the universe.

I return to myself.

"We should sue the urologist," my mom is saying.

"Stop it, Samia."

"They should have caught this *months* ago. It's total negli-gence!" She licks her finger and pages through the notebook, scratching her pen under lines of words.

"We are not suing," he says, voice raised. He sinks back into his pillow, moaning, and she reluctantly tosses the notebook down onto the nightstand.

"Okay," she says.

"Okay," he says, wiping his eyes.

He turns to me. "We think it might be a good idea for you to spend this semester living at home rather than on campus. I know it's your last semester, but it would give us more time together."

I blink. "Of course. I'll be here." Still holding my hand, he set-tles into the pillow, shuts his eyes, and tumbles into the welcome abyss of sleep.

After my dad's diagnosis, time disintegrates.

I feed myself, attend doctor appointments, type at my computer.

I drunkenly call my best friend, Serry, and she tries to soothe my sobs.

Hours melt into days and when I blink, they've gone.

We attend five doctor appointments in five days, and my mom finds an oncologist at MD Anderson who develops a treatment plan: six rounds of Docetaxel chemotherapy, three weeks apart, then thirty rounds of radiation, then a lifetime of hormone therapy.

Within a week of starting Docetaxel, his hair begins to fall out. Two weeks, and he's lost fifteen pounds. Then it's Christmas and we're at a family friend's celebration and I'm wondering if this will be our last Christmas together. When we sing carols, my dad's and my sacred tradition, I see his lip quivering, composure breaking, and I fight to keep my face blank, to paste on a smile. By the last song, I spring up and excuse myself to the bathroom, where I crawl onto the floor, wrap myself in a long red towel, and hug myself while I cry.

My dad and I begin to go on long, aimless drives. We take to the winding roads of the Hill Country and turn up the volume to sing along with his favorite bands: Talking Heads, the Cars, Ladytron. Outside of downtown Austin, the bones of rural Texas begin to poke through: we pass stretches of bone-dry drought-ridden fields; pebbled country roads; abandoned, buckling tennis courts. The dusty smell of the country slips through cracks in the car's windows and seeps into our skin. When I return home, I wait until bedtime to finally rinse it off.

When we spend time together, I tell myself to cherish each moment. But the seconds ticking by masquerade for normalcy, and the longer we go, the tighter the coils of dread twist in my chest. How can I cherish moments so fleeting they vanish long before I'm able to feel them?

In a parallel universe, I would have been looking forward to my last semester of college and preparing for the upcoming astronomy conference.

In a parallel universe, my dad would live forever.

Weeks pass.

The weather dangles between the clutches of winter and the promise of spring, lurching from ice storms to sun-soaked afternoons. I, too, exist in the in-between, half consumed by my dad's illness, half latching on to astrophysics, the latter mostly because if I surrender entirely to my terror, I know that I will lose all ability to function, that I will collapse onto my bed and opt to sleep and sleep and sleep until "day" and "night" lose their meaning, mixed up together into one.

I finish my research poster for the astronomy conference. At FedEx, the pale-faced, freckled attendant asks if I'm okay when I press the blue Print button and begin to cry, and I tell him that I'm fine, just fine. Gingerly, he takes it from my hands and laminates all 48 × 60 inches of it, telling me that this is the biggest poster he has ever seen and oh my heavens, am I really a physicist, have I discovered aliens, and what is a girl like me doing in physics anyway?

I excuse myself to wait outside in the rain, which is falling in icy sheets from the sky, while he lays the poster down onto the tiles and rolls it up into a thick spiral.

In early January, the oncologist gives us flimsy black-and-white slides that show five cloudy spots—tumors—where the cancer has metastasized: the left hip, right bottom rib, and L5, the lowest vertebra of his spine. He warns that cancer treatment will cause his bones to be more fragile, that they can now splinter easily.

My nightmares oscillate between grisly deaths—my dad bellowing in pain on the operating table, hellhounds clawing at him from the inside—and what I suspect are premonitions of the upcoming conference: I forget all physics, wander through the hallways naked, scientists lining the walls ordering me to leave.

All of a sudden, it's January 17 and I'm bound for Florida. At the airport, before I leave, my dad whispers in my ear, "Don't worry about me. Just focus on your poster session."

It is only later, when I'm sitting in a blue-ribbed seat 36,000 feet in the air, watching rivers and cities and canyons unspooling below, that I wish I had said, "It's not so much that I won't. It's that I can't."

The conference hotel is a maelstrom of tourists, scientists, students, spectators, and company recruiters. After checking in, I walk past a ten-foot-tall purple banner with 2016 AMERICAN ASTRONOMICAL SOCIETY stenciled in gold to the left of a row of elevators and into a vast atrium.

Foot traffic clogs the hallway to the conference rooms, astronomers elbow to elbow as they meander through. Dozens of booths line the atrium, some disbursing space-themed paraphernalia, a few selling industry jobs in the space sector, one offering a virtual-reality tour around the solar system. NASA's booth draws the largest crowd, attendees fawning over a human-size rocket and the accompanying shuttle flanking the table. Tables are interspersed between the stands, designed for conference-goers to mingle over drinks or a bite of food. And then, to the left, are rows and rows of posters, each one tacked onto an easel.

In the glare of fluorescent lights, the male scientists look identical and endless, each one hastily clad in a partially ironed suit, pale-faced, with tufted beards and the hint of a frown, as though they've been assembled in a factory and then spit out into this room. The few women stand out, their dresses flashing and their heels clacking loudly against the floor.

I step through the crowd quickly, avoiding the stares behind me. Amid the clamorous baritone and mud-colored loafers comes the overpowering knowledge that I am so out of place, so bereft of knowledge and skill, that those doubting voices over the years were, in fact, correct; that I might as well board another aircraft and return home.

I hold my breath until I step through a set of double-paned doors. Heaving, I take the elevator up four floors and finally make it to my room, where I kick off my shoes and collapse, trembling, on the percale sheets.

I drag myself into a seated position and attempt to recite my script before dinner, yet scenes from my nightmares swim into my mind, clogging my thoughts. I try to convince myself that they're not portents of failure or death; still, dread coils around my insides and squeezes until I can hardly breathe.

At seven o'clock, when the air turns chilly and the sun dips just below the horizon, I meet a fellow astronomy undergraduate, Haley, and two middle-aged astronomers in front of High Tide Harry's for dinner. The sandy-haired, grinning one in striped, red flannel, Adam, is at UC Santa Barbara; the other one, his friend, Rob, dark-haired, observant, with a rakish smile, is from a large East Coast university. Haley knows Adam from research they've worked on together and invited me to join them.

We get a table in the middle of the main room and order a round of drinks: an IPA and tequila shot for Rob, lagers for the rest of us. I try not to fidget with my hair and betray my nerves and instead clamp my hands together under the table.

"Is this your first astronomy conference?" Adam asks pleasantly, and Haley and I nod. Haley is an Air Force veteran, five years older than me, and speaks with the two men easily.

"Yes, but we're so excited to share our work! How many have you two attended?"

"Hmm, over a dozen, I think," says Adam. Rob nods, downs his IPA in a single gulp, and waves the waitress over, a pretty young woman with long jet-black hair.

He gives her an appreciative once-over and then asks, "What's your name?"

"Becca," she says. "Can I help you?"

"Well, hello there, Miss Becca. Can I get another one of these?" He gestures at the empty glass and then watches her backside as she strides away.

"Do you two enjoy teaching?" Haley asks.

"Absolutely," Adam says, nodding vigorously. "It's one of the best parts of being a scientist."

Rob looks at him skeptically.

"I wouldn't say it's the best part, but it's okay. Although there are perks," he admits.

"Like what?"

He grins devilishly and takes the shot. Lips smacking, he says, "I swear to God I have women—students!—falling head over heels for me." He picks up his empty beer glass and frowns. As though on cue, Becca arrives and hands over his second IPA.

"A couple of girls showed up at my office hours to say they're into me," he continues between sips. "They have these wild fantasies, you know, and it just makes teaching so much more enjoyable." He smirks.

Wild disgust coursing through me, I pick up my lager and force it down. His unbounded arrogance—the possessive way he looks at the waitress; the light in which he casts his female students: thirsty, lustful, wanting; his nonchalance when sharing these stories—it makes me sick.

"Hey, Miss Becca!" He flags her over once more, and she narrows her eyes before approaching the table for a third time. Adam is looking around the restaurant as though searching for a distraction, avoiding eye contact with all three of us. I fight against the urge to shout at her, to keep her away from Rob's preying stare.

"Yes?" she asks.

"Well, I was just wondering," he says, "can I get your number?"

"No, thank you. I don't date customers."

"So like"—Rob winks—"what if it isn't a date?"

She rolls her eyes. "I'm fine. Can I get you anything else?"

"Hey, I'm a great guy. Adam, tell her."

"I think you've probably had enough to drink, yeah?" Rob shakes his head unevenly, and Adam speaks directly to Becca for the first time. "Can we get the check?"

She gives a curt nod before disappearing, and Rob squirms.

"These *women* are so high-maintenance," he huffs, "it's unbelievable."

In the car ride on the way back to the hotel I shut my eyes. Visions swim into my mind: my physics professor, who showered our class with jokes about prostitutes and infidelity; that astronomer from science camp; Ahmad, who wrapped his hands around my neck and slammed me into a wall when he learned I'd been texting Steven, a member of my research group.

Have they ever been held accountable? How many women have had their math and science dreams snuffed out by men like them? Part of me wishes those women had stayed the course, and part of me realizes how dangerous that would have been—how impossible.

When I crawl into bed, my phone vibrates twice: a text message and a message request on Facebook.

HALEY: Hey, weird night. u ok?

The blinking Facebook message is from Adam.

ADAM: Hi Sarafina, I want to reach out and apologize on behalf of Rob. He was pretty inappropriate. He's not normally that bad. Too much alcohol, I think haha. Let me know that you made it home safe and sorry again.

I curl onto my side and stick the phone under my pillow, stomach churning. If Adam thought it was so inappropriate, why didn't he say something? Did he think Becca was asking for it? That Haley and I didn't care?

I fall asleep playing the night's events over and over again in my mind and hating that this is what I'm thinking of—not my dad, not my poster session, but this.

The first thing I do in the morning is down a cup of coffee and immediately throw it up. Twice, I rinse my mouth with toothpaste and soap my skin, but the acrid scent remains.

By the time I have made my way downstairs, clutching my rolled-up poster in one hand and my scribbled script in the other, I have imagined seven different ways this could be a disaster. At the sound of every man's voice, I pray that Adam and Rob haven't found me here, that I can make it through this poster session without facing them.

Fluorescent lights glint off the freshly polished ceramic tiles lining the ground of the convention center. I wonder what they will look like with spit-up brown sludge sprayed onto them.

With trembling fingers, I pin my poster to an easel in the center of the fray and mutter the script under my breath. Footfalls echo along the tiles, and when I look up, a scruffy-faced man, probably in his mid-thirties, is standing directly in front of me wearing a lopsided grin, cargo shorts, and tortoise-rimmed glasses.

"Hi," he says, "I'm Aaron, a professor at UC Berkeley. I saw that you're one of our graduate student applicants this cycle! Would you mind telling me a bit about your research?"

Nausea. I swallow the taste of vomit, inhale long, deep breaths.

"A-absolutely," I stammer, "it's great to meet you."

I try to reassure myself that I have prepared for countless hours. Long nights spent hunched over my laptop, sunbaked afternoons scribbling at Dr. Wheeler's blackboard, simulation after simulation after simulation.

And then, a vision of the night sky swims into my mind, a trillion pinpricks of light dangling in the darkness. For a split second, the rush of the moment burrows into me, as though the fog of fear winding through my mind has lifted and I can see everything clearly: I can do this, I am capable and worthy, I am right where I need to be.

My breath steadies.

"Last summer," I say, "I worked at Harvard's Center for Astrophysics modeling supernovae to try to determine how much, if any, hydrogen is left behind when the star explodes." It was the summer after breaking up with Ahmad, after I'd finally decided, with Dr. Wheeler's encouragement, to apply to research internships for *myself*—and, to my utter shock, made it all the way to Harvard.

I spent three months in Boston. Twelve weeks of losing myself in supernova explosions and exploring another city far, far away from home. I spent my days flitting from one coffee shop to the next, never staying long enough to settle in thanks to the old, worn fear that Ahmad would somehow follow me there and find me.

Through the many intertwining circles of young people in Boston, I made a new friend, Serry, who was conducting biology research at MIT. Together, we'd work late into the night, and with Moscow Mules and pizza, we bonded over our love for science—and feeling out of place as women of color in male-dominated fields.

"Sometimes," I confessed to her one night after a particularly grueling day of research, "I wonder if all of this is worth it."

Yet we pulled each other through the summer. Studying with her was easier than studying alone, and she inspired me to redouble my efforts.

"We've got this," she said to me, grinning. "We have each other."

Under my research advisor's guidance and with Serry's encouragement, I ran hundreds of simulations exploding single and binary star systems. And after a while, to my great surprise, they began to make sense. During my final presentation to the Harvard astronomy faculty and the other internship students, in between fighting waves of anxiety, the knowledge settled deep into my core that I understood more than I was giving myself credit for.

Maybe, I thought to myself at the end of my talk, I do belong here.

It is the same realization that occurs to me now.

"You mean," says Aaron, breaking my reverie, "we might be misclassifying supernovae?"

I blink and smile. "Bingo. The amount of hydrogen could imply something about the formation mechanism of the supernova and its progenitor—for example, is the original star a supermassive star or in a binary system?"

"Huh." He tilts his head and pushes his glasses up his nose. "Tell me more."

"Well," I say, "we toy around with several parameters. Does the star rotate? How much? Does it have intense winds stripping the surface?"

"So how do they impact the timescale to explosion?"

A viselike panic squeezes my chest. How would Dr. Wheeler answer this question? In my mind, I conjure up the equations lining his chalkboard, the bones of derivations we solved together.

Remember, he once said, *when you're thinking about stars, you have to consider the whole system.*

Line by line, I imagine solving the luminosity equation, sliding variables and exponents into place. From the depths of my mind surfaces a memory of my quantum mechanics course, when we learned about Hamiltonians and spectra, and all of a sudden, I know how to answer.

I point to a graph in the center of the poster that superimposes spectra from various types of stars onto the same axis. One giant red line marks the hydrogen feature.

"Effects like winds and rotation strip the surface of the star, effectively determining the type of subsequent supernova. This might impact the time to explosion, but it also might not!"

The shadow of a smile.

"Based on our simulations," I say, "we find traces of hydrogen in stars we never would have previously considered."

Aaron nods, looking impressed. "Excellent work. I look forward to seeing the paper."

He then asks whether I enjoy my research and what fields of

astronomy I'm interested in. Minutes tick by as the conversation shifts away from raw science, and before I know it half an hour has passed and he's telling me that I did great work, that he can't wait to review my UC Berkeley graduate school application.

When he strides away, an enormous weight lifts from my chest. Until now, I realize, I have been so obsessed with *being*: perfect, brilliant, astute, observant; that I have missed the whole point of *doing*—Mikan's observatory; Dr. Wheeler's chalkboard; the McDonald Observatory; these moments when I lose myself to the stars and feel the science winding deep into my bones.

This. This is why I do it.

By the time the poster session is over, I have recited my script to eight faculty members and three postdocs from various institutions. My mouth is so dry it tastes of paper, and the blisters from my heels have burst long ago. Limping, I make it to a water fountain in the main hallway and swallow several long drinks.

Relief courses through me. The morning went far better than I'd dared to dream. A professor from Caltech said that my simulations were both useful and relevant, and another, from UC Santa Barbara, urged me to apply to their department. Only the University of Arizona research scientist looked at me skeptically, only he said that my simulations need to be reworked.

In fifteen minutes is the cosmology keynote, "A New Universe of Discoveries," in Room C. To kill time, I find a bench in the back of the atrium and slide into it, weary but excited.

There is only one person on the planet with whom I want to share this moment.

I pull out my phone and dial.

"Dad?"

"Hey honey," he says, "hang on a second." He covers the speaker to say something I can't hear, and then says, "Sorry, that was the nurse—"

"The nurse?" Alarm descends down my spine.

"Yeah," he says, "I'm at the hospital getting chemo right now."

To the left of the enormous conference sign, an astronomer is talking urgently to his colleague, flapping his hands and pointing at a graph. Others join in, and their laughter rings across the room.

All of a sudden, the atrium is too hot, too full, and all of these people with their starched shirts and flimsy posters are too loud.

"Shit." I exhale sharply. "I'm sorry. I don't want to interrupt."

"No, it's fine, honey."

I lean against the beams of wood poking into my back. "Are you okay?"

"Oh yeah," he says, "this is my routine now. Chemo on Tuesdays and Thursdays every week!"

"Every week," I echo.

I picture a chemotherapy pump sticking into his arm, injecting toxins into his feeble body. Even though I am one thousand miles away I can smell the hospital stench, the nauseating mix of antiseptic and pus and blood.

I think of a sterile, clinical room, where cancer patients sit beside drip stands and IV bags in front of an empty wall, some attempting to make conversation, others wholly lifeless, the overwhelming sense of sickness and death hanging in the air.

"Enough about me," I hear him saying, "tell me about the poster session!"

The sensation of untethering digs into me: for a split second, I am no longer here; I am nowhere, hanging as though by a thread in the endless blackness of a crushing void. I remember none of my discussions, no physics: nothing about where I am, what I have accomplished, means anything at all.

"Honey?"

"Yeah, sorry," I say, "I'm here." I shut my eyes against the glinting tiles and allow my head to fall back. "It, um, went well."

"Yeah? Did you get some good questions? Chat with anyone you wanted to meet?"

Drop by drop, I can feel tears collecting behind my eyelids. I have to get out of this room.

"Actually," I interrupt, "can we talk about this later? The keynote is starting."

"Sure, of course," he says. "Have a good time today, honey. Go learn something!"

The phone slips from my fingertips and falls onto my lap. I try to blink back the tears now coursing down my face and blindly grab my poster before sprinting down the hallway, into the elevator, and then finally to my room, where I throw the poster onto the sheets. It tears when I collapse.

I should have stayed with my dad. I should have gone with him to his chemotherapy appointments; I should have noticed his illness sooner; I should have helped him. I want to choke on the shame, the knowledge that I am *selfish* and *wrong* for coming here, that nothing is paramount to his well-being, his health, his life.

The overwhelming urge to fly back to Austin and sit in that chemotherapy chair, to stick the IV into *my* veins and suffer the dizziness and nausea and hair loss and fatigue instead of him, flows into me.

And then, I get a text from my dad, a simple one-liner:

I'm so proud of you, Little One.

Months roll past. Dad finishes chemotherapy, begins Eligard hormone therapy, becomes vegan, quits alcohol. In our tiny apartment my mom and I watch him change, so much so fast, a fifty-eight-year-old emerging from the shadow of corporate America and unfurling into someone new.

He stops working all the time. Life is too short, he tells me. He begins to organize and stack his records, making piles by year and then by artist. Hundreds upon hundreds of disks strewn around our living room, and he is happier than I can remember.

When we get coffee together, we now speak about death.

"I know it's silly," I say, "but I think that death is my greatest fear."

Hands knocking together, he intertwines his fingers and gives me the shadow of a smile.

"What's to be afraid of? It comes for all of us."

Late one night, I confess that I don't want him to leave me.

"I don't know what I'll do without you," I say, trembling, wiping my eyes and curling into his chest.

"I'll always be with you," he says, "even after I'm gone. Right here on your shoulder." He taps my left side and smiles at me sadly.

Most nights I spend working on physics problem sets at the round kitchen table. The knowledge that this is my last semester, that graduation is in just two months, keeps me steady, focused.

To my surprise, I love this semester's physics lineup: quantum chemistry and particle physics. The best moments are when formulas slide together seamlessly, and for a few precious minutes I can see the problem a million different ways, from the foundational first principles all the way to how they play out in the universe. During those moments my heart throbs: it is all so beautiful.

On a warm evening in March, when fireflies dance and the twilight gets longer, my phone rings.

"Hi, this is Mariska from UC Berkeley. Is this Sarafina?"

Through the window, I spot a swallow swooping down from the sky and whistling into the night. Tree branches swing lightly in the breeze. My heart begins to pound.

"Yes, speaking! How are you?" I slide the glass door open and walk onto the balcony. All above me, the stars glitter.

"I'm great," she says. "I wanted to call you to personally let you know that you've been admitted to UC Berkeley's graduate program in astronomy!"

The world stops.

I wrap my hands around the cold metal banister lining the balcony. A roaring fills my head, as though I have been dunked underwater and there is no sound but the blood rushing to my skull.

"Congratulations," I hear her say. "You're going to be an astrophysicist!"

Everything is glowing: the fireflies, the stars, the headlights out in the distance. The breeze tickles my face; I feel warm drops lining my skin and realize I am crying.

"—send along materials about salary, important dates, healthcare, stuff like that," she is saying.

I don't know which words spill out of my mouth, but she is laughing and telling me you're welcome, that I'm a great fit for the department. Moments later, we hang up.

The door slides open once more, and my dad walks out onto the balcony.

"I saw you crying," he says. "Are you okay, honey?"

I turn toward him, tears still spilling down my face. In the starlight, they glitter too.

"That was Berkeley. I got in."

The roar of his laughter, the amazement in his eyes, the strength he finds to wrap me up in a bear hug, the skin of our cheeks pressed together and the salty taste of his tears mixing with mine.

"I knew it," he says, grinning ear to ear.

"I didn't."

"I know," he says, laughing, "but isn't this proof? You are more than capable."

Under the canopy of stars, he hugs me once more. I want to bottle his love and pride and admiration, to stopper it in my pocket, to freeze this moment in time.

"Thank you for being here," I say.

"Always."

CHAPTER X

FUTURE OF THE UNIVERSE

Everything in the universe will die—even the universe itself. The timescales for their deaths vary—atoms might decay in just a few minutes, a dog might live for a decade and a half, humans might live for eighty years, stars might live for a few hundred million. The universe itself might live for hundreds of billions of years, a profoundly incomprehensible timescale. Yet we know, with some certainty, that sometime in the far distant future, many eons from now, the cosmos will be no more.

The ultimate fate of the universe is determined by a battle of cosmic forces: gravity, which binds stuff together, and dark energy, which stretches spacetime apart. The more *stuff* there is in the universe, the greater the force of gravity, as it binds more and more mass together. On the other hand, the more space there is, the weaker the force of gravity—the farther away things are, the harder it is to knit them together.

With one caveat: if dark energy is not constant, but instead changes with time, growing ever stronger or falling progressively weaker, then the battle must be recalibrated.

Everything in the universe—all of the matter, like stars and galaxies and dark matter, and all of the energy, including dark energy—shapes its geometry. Einstein's theory of special relativity ($E = mc^2$) demonstrated that matter m and energy E are linked—in fact, they're one and the same when we include the speed of light, c.

Both matter, which warps spacetime, and dark energy, which stretches spacetime, physically shape, distorting and bending, the curvature of the universe. To understand the fate of the universe, we must understand its shape.

There are a few possibilities. An open universe, with negative curvature and infinite volume, is shaped like a horse's saddle or a potato chip. In this geometry, the universe continues to expand forever, and galaxies separate more and more quickly with time, until all structure is ripped apart.

A closed universe, or one with positive curvature, is shaped like a ball. Unlike the open universe, the universe eventually collapses back in on itself as gravity overcomes the force of dark energy in a Big Crunch. At the end of time, everything gets so squished together that the temperature skyrockets and the universe ignites in a massive explosion.

The last geometric configuration is a flat universe—one shaped like an infinite piece of paper. Dark energy forever propels this universe's expansion, although galaxies separate more and more slowly with time.

As far as we know, dark energy is a fundamental, constant, pervasive property of space. Thus, as space continues to expand—and accelerate!—there will be more and more dark energy, and at some point in the far distant future, there will eventually be more dark energy than matter.

At that point, dark energy will be the victor in the cosmic battle for the fate of the universe.

Based on our current measurements, we believe that the universe is flat. And because of that flatness, we anticipate the universe's fate to be dark and bleak.

In the stelliferous era, the age we're currently in, where stars abound and light fills the cosmos, stars grow and die, seeding the interstellar medium with elements from their cores. But if dark energy continues to accelerate the expansion of the universe, we expect the space between galaxies to grow, ultimately becoming so vast that galaxies are carried immeasurably far away from one another. The temperature of the cosmos will plummet and starlight will dim until no visible light reaches us at all.

In 10^{14} years, the gas-fueling stellar nurseries will be exhausted, and the Age of Stars will be over. In 10^{15} years, stars will go extinct and we'll be plunged into darkness, heralding the onset of the Degenerate Era.

Black holes will feed upon leftover stellar relics, like degenerate neutron stars and white dwarfs, until nothing but black holes remain, and the universe will be frigid, empty, and dark, as black holes themselves evaporate.

In this final stage, the Dark Era, solitary elementary particles, neutrinos, and low-energy photons will be the only remaining "things" in the universe. They'll be sprinkled through spacetime, until eventually all that is left is nothingness: a desolate, cold, and unforgiving shell of a universe.

Months slide by like honey from a jar.

Spring showers spill out of the sky and a group of professors asks me to deliver the student commencement speech to all of the math and science majors of the College of Natural Sciences, thrilling my parents. Then temperatures rocket into the hundreds and the lake beds run dry, and all of a sudden I am finished with college. On the first day of over one hundred and ten degrees, my dad's oncologist says that his cancer has progressed and recommends a second round of chemotherapy, and I call UC Berkeley to defer graduate school for one year.

Come autumn, I move into a one-bedroom apartment, accept a mundane job doing market research for hedge funds, and winds push my bike down the road, over the Fifth Street overpass and along the sidewalk, until I reach the enormous granite office building at 301 Congress. When I come back down the elevator and exit through the revolving glass doors, I blink and once again it is winter and ice glazes the streets.

At the beginning of January, after months of discussions, my therapist convinces me to get a dog to help with my anxiety. I want a puppy, I tell her, who will know my dad while he is sick—and after he is gone.

Comet is small, just four months old, with one ear that points up to the sky and another that flops down toward the ground. Already, I think, he intuits when I'm on the verge of a panic attack. When my breathing grows shallow, he places his head on my lap and stays there, eyes closed, until the storm passes.

One month after I pick up Comet, when the chill snakes through the cracks and into my lungs, I slip on a pair of gloves and a hand-me-down wool sweater from my dad and drive to Elizabeth Street Café. Texas wisteria cascades down branches flanking the sidewalk. Half a dozen purple buds have fallen, trampled in ice and dirt on the ground.

Up on the patio, my dad sits under a stainless-steel heat lamp flaring in amber, flipping through a binder of papers. Tugging the wool closer to my skin, I slide into a chair across the table.

He looks up, slides his chair out to greet me, and a weight drops through me. He is thin—far too thin. Reaching out to hug him, I wrap my arms around him easily, one arm resting on top of the other, and fear crawls into my chest: until two months ago, I couldn't touch my hands around his torso.

All around me, I hear the clamor of honking trucks, a shouting cyclist, and a refurbished Mustang playing loud music through its stereo. When I sit back into the aluminum chair, the sounds rattle my bones.

Unbelievable that this is the same man I've known my whole life.

"How was work?" He sits back down and sips a glass of ice water.

"Extremely dull," I say. "But you already know that."

He huffs a laugh, deftly turning the binder over and sliding it behind the water pitcher.

"Yeah, well, that's the financial sector."

"What's that?" I point to the binder.

"Oh, we'll talk about it after dinner."

I take in the way he is hunched over his plate, the way he wiped his eyes before I sat down. Fear lurches through me.

"What's wrong?"

"Little one," he says, "we can eat first."

Something is wrong.

"I can't—" I whisper.

Sighing, he meets my gaze, nods, and pulls the binder back out.

"All right," he says wearily. "I understand. So, as you know, I've been getting chemotherapy the last few weeks. Overall, I've had a pretty good reaction to it."

I nod furiously. "I know."

"Despite the hair loss, which I think I'm rocking." He winks. A ghost of a smile. "You know, by the way, my mom—your grandmother, Grannell—had a terrible reaction to chemo."

"Which time?"

"Well, back in the '80s, when she had ovarian cancer, she was on carboplatin for nearly a year. Nasty stuff," he says. "She had major nerve issues. Completely lost feeling in her legs and hands."

How many times as a young girl did I bake cookies with Grannell and watch her flex her hands, as though her fingers had fallen asleep?

"I didn't know."

"That wasn't the worst of it," he continues. "Her nails fell out, she threw up all the time . . . at one point, she told me she'd rather die than have to take that drug. She didn't take it when she was diagnosed with pancreatic cancer," he says. "She didn't want it."

I remember Grannell during those last few months, when she was so frail she could barely sit up. All of her passion for artistry, her years of work in design and architecture, her job as a female computer scientist: all of it was flattened into this waif of a human, who at the end of her days was carried into the wind.

A waitress sidles over to take our order; my dad smiles to let her know that we need a few more minutes, that we'll wave her over when we're ready. The smell of grease and fried rice is overpowering.

"Anyway," he says, "when I started the second round of chemo, Dr. Corn recommended that I do genetic testing."

"What is that?"

"It's actually very cool," he says, flashing a fleeting grin. "You'll appreciate the science. They drew some blood and then tested it in the lab for genetic mutations—"

"Why?" I interrupt.

"Two reasons. One, my cancer was extremely aggressive, which is a potential sign of a genetic mutation."

"Which is why they didn't find it sooner?"

"Exactly." He sighs, and now the tears have leaked down his cheeks and he is wiping his face and my heart rockets down my spine.

"The second reason is because some genetic mutations are hereditary," he says, now barely above a whisper. "Given our family history, they thought I was a good candidate."

Suddenly, the air is clear and I can see the outlines on all of the wisteria petals, the rungs of the bike wheels cycling past us, the pedestrian halfway across the bridge. A ringing starts in my ears; it clangs against my dad's words.

He meets my eyes.

"I came back positive for something called the B-R-C-A mutation," he says, spelling it out. "More than likely, I inherited it from my mom—your grandmother."

A throbbing starts in my temple, pounding to the beat of the fear pulsing in my chest.

Matter-of-factly, he continues. "The BRCA mutation is what's called a germline mutation," he says, "where the carrier has a fifty-fifty chance of passing it along to their child."

Like the puzzles we used to build together, his words fall into place, one by one, and suddenly I understand; I see it all, the whole picture that has been in front of me all this time.

"I might have it, you mean."

His face glitters with tears.

"Yes," he says simply. "You might."

The next morning a thick gray curtain of haze hangs from the sky and my dad and I drive to Texas Oncology. The air is charged and the wind gusts across our faces as we walk from the parking lot, through a set of glass doors and up to a half-moon check-in desk, and I feel the weight of the coming storm along my shoulders, crushing against my chest.

The waiting room is full of patients. The youngest are middle-aged women, one with a red-and-white-checkered kerchief tied around her head and another with no eyebrows; then there is the silver-haired man hunched over his cane, a tennis ball stuck to the end, shuffling to his chair; the couple in the corner each have a dead look in their eyes, staring at the wall together; and then there is my dad, and there is me.

Queasy, fighting goosebumps, I walk up to the front desk with my dad.

"Checking in?" the attendant asks my dad. Mute, he shakes his head and wraps his arm around me.

"Checking in for her. Sarafina El-Badry Nance."

The attendant blinks and then expertly slides on a professional face. *Yes, of course, my mistake, let me check you in here, please fill out this paperwork and sit over there, we will be with you shortly*—I have cotton in my ears and the words do not make sense, they rebound off me and hang somewhere in the space between us.

Minutes pass, and then a dark-haired woman in her mid-thirties emerges from a back room and calls my name. Together, my dad and I follow her to a round table, and we all sit down.

"My name is Katie. I'm a genetic counselor here," she says. "Your dad's oncologist told me that you're a scientist, right?" She asks me directly, presumably to put me at ease, but my muscles lock up and my answering nod is rigid.

"Since your dad came back positive for the BRCA mutation, I'd love to use this meeting to share some details with you about this specific mutation." She says all of this slowly, her lilting voice calming.

I nod, unable to muster a response. She smiles softly.

"The BRCA mutation increases the risk of cancer and gives the carrier a fifty-fifty chance of passing it down from parent to child. All genders can inherit, although depending on your gender, the types of cancers you're most at risk for can vary. So, in your dad's case, for example, he has an elevated risk of prostate, melanoma, breast, and pancreatic cancers."

As she speaks, my dad again wraps his arm around me. I lean into his scent, letting it soothe me.

"In your case," she continues, "if you carry the mutation, you would be at increased risk for breast, ovarian, melanoma, and pancreatic cancers."

I see her speak and feel my dad go rigid, but I feel nothing, no twist in my guts nor grip on my chest; I am numb.

"When we draw your blood, we'll test it for the full spectrum of genetic mutations. And then, if anything comes back positive, we'll run models based on your family history to determine your lifetime risk of these types of cancer. Everything make sense so far?"

I hear the words—they make sense insofar as I understand that they mean something, but their implications, their impact on the rest of my life . . . no, no I will not let myself get that far.

"I—" The word gets stuck in my throat. "I think so."

"Again, we don't know if you'll come back positive for anything. But if you do"—she smiles kindly—"there are options."

"What . . . what sorts of options?" I ask, cringing at the trembling in my voice.

"Well, this is actually the cool science part—" She winks, knowing that I'm a scientist, and I force a tight-lipped smile.

"Genetic testing," she continues, "allows you to make informed decisions about your health. It equips you with knowledge to face illness head-on. In your dad's case, for example, his treatment plan might prioritize specific types of treatment, like immunotherapy, because BRCA tumors tend to be more responsive to treatment plans like this."

My dad straightens in his chair. "Wow," he says. "Even I didn't know that."

"I'm sure Dr. Corn will mention it in your next meeting." She winks again.

"But for those who haven't been diagnosed with cancer, we approach the BRCA mutation proactively. There are regular surveillance techniques, like MRIs, which would allow us to catch any cancer growth early, before it's progressed. And there are surgical options that some patients preventatively undergo to remove the organ that is at increased risk of tumor growth."

Cancer. Tumors. Surgery.

Inside my mind I am understanding, for the first time, that this could affect my entire life, that if I am positive for BRCA everything will change.

"Based on your family history," Katie says, "it is my professional opinion that you proceed with genetic testing."

My dad and I nod; we expected this.

My throat burning, I force the words out. "I agree. I'm ready."

"Wonderful." She shuffles her papers together and stands, gesturing toward the door to the lab. "Let's go get your blood drawn. You'll get the results somewhere between four to six weeks."

Weeks and weeks of waiting. Already I know that I will be unable to relax, that I will be waiting by my phone for a call.

"Okay," I say, "let's do it."

They draw twelve vials. Incredible that they can peer through drops of blood, unfold my DNA, and search, row by row, for missing or broken pieces. This science will reveal my cancer risk. My mortality.

I shuffle through the next few weeks in a stupor. The rigidity of my daytime schedule helps: wake up at seven, take Comet out, bike to work, type away at my computer, attend work happy hours, bike back home. Comet gets me outside, and when we are running together in the park across the street, I feel alive.

It is at night, when my work pauses and I have no structure, that the terror takes over, when I feel invisible claws wrapping around my heart and squeezing my chest, until I see black and blue spots and I sob so hard that I cannot breathe. Each time, Comet curls into me, places his head on my chest, my own weighted blanket of fur. His even breathing regulates mine, and although I suspect that my anxiety only amplifies his own, I am grateful all the same.

I start on prescription anxiety medication: Lexapro, once a day, every day; Xanax as needed for the panic attacks. My therapist diagnoses me with a severe case of generalized anxiety disorder and PTSD—common, she says, in victims of abuse.

I tell her I feel so helpless, so utterly worthless, that sometimes I think I don't deserve to exist. At this, she turns deeply sad.

"Everyone," she says, "deserves to exist."

She continues, "Abuse destroys one's sense of self. But together, we'll rebuild it."

At night, in the quiet moments, I attempt to make progress in my astronomy research. Dr. Wheeler has patiently agreed to work with me throughout my year off, even though I'm hovering in between college and graduate school. I have finished my simulations, already determined that the star Betelgeuse will not explode for another hundred thousand years—all that is left is to write the damn paper.

Some nights, the solidity of the numbers is a soothing distraction. They have no feelings, no life cycle; they exist independently of the events unfolding in my little pocket of the universe. The thought is comforting: *none of this matters.*

Other nights, I am so numb, so exhausted, that I crawl into bed with Comet the moment I return home.

On the day that marks four full weeks since my genetic testing appointment, I am sitting in a corner room at the office, mindlessly typing up a brief on mattress retailers, the glass and polished stone of skyscrapers outside the floor-to-ceiling windows glinting against the midday sun, when my phone rings.

"Hi, this is Katie from Texas Oncology. Do you have a moment to chat?"

Immediately, black spots speckle my vision and I fall back into my chair, the walls blinking in and out.

"Yeah, of course."

"The results for your genetic test have come through," she says, and I know what is coming, I can hear the overwhelming sense of urgency in her tone. The sadness.

"Unfortunately," she says, "you came back positive for the BRCA-2 genetic mutation."

Her voice is loud and soft all at once; my ears are raw from listening. Still, no emotion crests over me; I am simply numb.

"According to these models," she continues, "you have an 87 percent lifetime risk of breast cancer, a 30 percent lifetime risk of ovarian cancer, and an elevated risk of pancreatic and melanoma cancer."

I am here and I am nowhere, the same numbness that enveloped me when my dad told me he has cancer. The numbers she recites tick through me, but aside from a faint flicker of recognition, my brain is shutting down. I cannot hear any more, cannot process what any of this means.

"Since you're attending graduate school in the Bay Area," the voice on the phone says, "we recommend joining the BRCA center at UCSF and working with an oncologist who specializes in hereditary cancer. Specifically, we recommend Dr. Dhawan."

An oncologist. I will have my own oncologist, my own medical team. I, a presumably healthy twenty-three-year-old, will be a patient at a cancer center.

"Just remember," she says, "you have options."

My vision is rippling in front of me, the shapes of the buildings dancing and melting into one another.

"Sure," I say.

She continues speaking, and then we hang up, and my phone slips through my fingers and falls to the table. Bucking against the nausea roiling in my stomach, I pick it up and dial.

"Hi honey," my dad answers.

I hate that I have to tell him this, hate that I am the one to burden him—

"Hey, Daddy, Texas Oncology just called."

A sharp inhale.

"I . . . I got the news." Something inside me fractures, and I know that it's irrevocably broken. Tears slide down my face.

"I'm positive. For BRCA-2. Like you. I have it too."

"Oh, little one," he says, voice cracking. "I'm so sorry." I hear him sniffle, how he fights against the tears. My throat tightens so quickly it hurts.

I cannot handle this, cannot be the one to cause him pain, cannot shoulder this burden, cannot bear this stress every day for the rest of my life—I cannot catch my breath; I am heaving, gasping, whimpering against my tears.

I press the phone to my ear, and with my other hand I fumble through my backpack, searching-searching-searching for the orange-and-white bottle of Xanax.

I pop one of the almond-shaped white pills into my mouth and slug back a sip of water, praying that it releases into my bloodstream soon, praying that it takes this *panic* away—

"I'm here with you, honey," he says. "You're not alone."

Something in his voice gives me pause, a note of strength or desperation or both. I gasp for air and finally feel like I can inhale a breath, hold on to that precious oxygen before it evaporates.

He waits for the tears to slow, for an impossible amount of time, until my breathing finds a rhythm once more.

"You know now," he says quietly. "Long before anything happens, you know. That takes strength, Sarafina."

I hold on to that strength, let it envelop me for a moment.

"This—this gives you options. And you did that for yourself. You did. Nobody else."

I hiccup a sob, letting his words wash over me.

Then he says, almost too quietly for me to hear, "Unlike me, or your grandmother, you can be proactive. You get to decide."

A faint image of Grannell flickers in my mind, determined and unwavering, bent over me just days before she died. *Make good choices.*

"Make good choices," I whisper.

"You're on the right path, little one. I know it's hard, but I'm right here with you," he says. "Every step of the way."

I make one more call.

Fifteen minutes later, my friend Taylor, a dark-haired man around my age, with gentle eyes and a kind smile, picks me up from the office in his black 4Runner. A coworker recently introduced us, and we immediately bonded over his phone's wallpaper of the solar system. Nights of *Battlestar Galactica* reruns and conversations about supernovae form the bedrock of our friendship, and we laugh easily together about how similar we are—two nerdy dog lovers who enjoy geeking out about the universe. But above all, it's his compassion that drives me to call him now.

He asks no questions when I collapse into the passenger seat. Instead, he simply reaches over, rubs my back, and drives.

Only when we have turned away from downtown, when the office is no longer in sight, do I allow my tears to fall in earnest.

In silence, we drive across the Pennybacker Bridge, over rolling hills, across streams of water, away, away, away.

He doesn't press me. But the farther we go, the more I'm struck by his companionable silence, a quiet that allows me to just *be*.

Finally, when we've slowed to a stop by one of those flickering streams, I muster up the courage to tell him.

"I got the news today that I carry the same genetic mutation as my dad," I say softly. "It means . . . it means I'm at extremely high risk of cancer."

He meets my gaze, eyes soft, and wraps his arms around me.

"I'm here. For whatever you need, I'm here," he says, ever the stalwart friend.

Tears slide down my face and into his gray-blue Dri-FIT shirt. Still, he doesn't move away. Instead, he goes on. "In life, there are big, hard moments. This is one of those."

I nod, thinking he's finished, but he surprises me by continuing. "But we've got this. Together."

In August, when the drought-ridden Texas soil turns the color of rust, my parents tell me over dinner, just days after I quit my job to move to Berkeley, that they, too, are moving—to Mexico. *It's too expensive to stay here with everything going on,* my mom says, gesturing toward my dad. I just blink at her, numb to all of the changes.

My mom and I haven't spoken much about my diagnosis, although not for lack of her trying. Every time she attempts to bring it up, tears fill her eyes and concern creases her face, and I am so anxious, so unable to sit in my own discomfort, that I invent an excuse to end the conversation. No part of me is ready to engage with her, to comfort her. I don't realize that deep down, it's because no part of me is ready to confront my diagnosis at all.

Together, the three of us pack our things into Sharpie-labeled boxes. My dad, still in the throes of chemotherapy, tries to lift the heaviest objects—a leather couch, a mattress, a paneled wood dining table—and fails. He is feeble, too frail to lift anything heavier than a midsize box, and I see his frustration and shame mounting, until he finally collapses in a dining chair and closes his eyes.

I call eight friends to help. When they arrive, my dad sidles to the back of the apartment and looks on, wiping away the shadow of a tear.

Taylor, one of the friends who shows up to help, meets my eyes. He walks up to my dad, introduces himself, jokes that he'll have to visit my parents in Mexico, and then slips over to me and squeezes my shoulder.

"We can handle this," he says. He knows, understands that witnessing my dad like this breaks me.

To watch a man—not just any man, but *my dad*—once so strong, larger than life, be brought to this . . . it lights a flare of rage within me.

I do not ever want to succumb to this disease.

Just a few days later, Comet and I pile into my dad's blue Volkswagen sports car and drive across the country. When we cross the state border into Arizona, I am thrumming with anxiety, oscillating between feelings of dread and excitement—doing astrophysics full time is what I have been preparing for my entire life, and yet . . . I cannot help but chafe against the unreality of it all, how little this matters in the context of life and death.

Four sunsets pass, one at the peak of the Rocky Mountains, casting the mountains in shades of rose and amber; one across vast plains of ocher and clay; another in the stark white of the salt flats, glittering pinks reflecting off the brine; and the last in the crags of Lake Tahoe, wild forests and banks of boulders gobbling down the final rays of sunlight.

With each rotation of the Earth, my chest throbs a little harder with homesickness. I want to rewind the clock, to go back to the flitting days of summer when I played in the streams of West Texas and strolled under the stars at night.

Yes, those days weren't perfect, but at least back then my home—my *dad*—was safe.

I was safe.

I drive so long that my eyes blur, the jade of pine leaves blending with blazing yellow hills of wild oat and filaree. As Texas recedes farther and farther away, my anxiety mounts, until I reach the outskirts of the Bay Area and a crisp layer of sea salt wafts through the car and chases away any other thoughts.

For better or for worse, I've made it.

The Bay Area is absolutely nothing like Texas.

Cloudless skies tinted in hues of the clearest blue, flecked with clouds of ivory and rays of gold, span as far as the eye can see, casting the soaring skyscrapers of downtown San Francisco and tumbling hills edging Berkeley in sunshine.

The air smells of eucalyptus, soft brown-and-white trunks towering in between enormous redwoods and flowering palms. Every house is unique, tucked within a latticework of trees and moss, their front lawns often overrun with sprawling gardens of cacti and succulents, dashed with bursts of color from blooming dandelion and California poppies sighing in the breeze—this is a smorgasbord of plant life, utterly at odds with the dehydrated neutrals of the Texas Hill Country and its oppressive heat.

My first class, Radiation in Astrophysics, is at nine A.M. the following day. After hauling the boxes, bike, and my dog inside my new house, I collapse on the bed and fall asleep immediately.

The walls of my new bedroom close in as jagged knives spearing out in rows crack through the ceiling. In the center of the room is Ahmad, who towers over me as he pulls one of the kitchen knives into his hands and rests it against my throat—

I wake up to the sound of my own screaming. My new housemate is pounding on the door asking if I'm all right. I struggle to get the words out, to say that I'm okay, it was just a nightmare, a horrible nightmare . . .

The following morning, I blink back my exhaustion and untangle my bike from boxes of books and clothes. I want to revel in the excitement of the moment—how I am achieving what I have always dreamed of—but instead I feel nothing but anxiety.

I paw at my phone and text my dad.

I wish you were here with me.

On the ride to campus, a thick coastal fog blankets me and the bike in a damp chill that reaches all the way to my bones. I curse myself for not grabbing a jacket before leaving. Cold in August—so unlike the South.

Tendrils of dread twist in my stomach, and as I pedal by lawns of students and a soaring ivory clock tower, headfirst into the wind carrying the thrill of the new semester across the campus, I feel as though I am watching it all from afar, stuck behind a permanent and invisible pane of glass.

When I called UCSF to set up an appointment, the office manager told me in a sharp, crisp voice that I'm already behind, that I need to get an MRI as soon as possible to begin monitoring my breasts.

"But I'm only twenty-three," I pleaded. But they were gently firm.

The specter of illness hovering over me is suffocating; I feel like a mouse caught in a trap, waiting to be struck.

Pushing the thoughts to the back corner of my mind, I heave a breath.

Gleeful laughter soars through the air, sending a pang through my chest. If I'm honest with myself, I resent it—resent them—for that freedom.

Dreading the next hour and a half, I drag myself to the astronomy building, lock my bike away, and make my way to an older, wood-paneled classroom.

Slate-and-wood lab desks in rows of two line the room. Already, most of the seats are filled with half a dozen other astronomy graduate students and two physics students. It barely registers that I'm one of two women—I am so used to it by now that I've learned to ignore the faint thud of disappointment in my chest.

Just as I slide out my notebook and ballpoint pen, Aaron, our professor, strides in and dumps a stack of papers onto the table.

"I believe this class is most, if not all, first-year graduate students," he says. "So, to all of you: Welcome to Berkeley!"

When I look around, I see that most of us look equally nervous, save for the clean-shaven student sitting at the front. Aaron gives us a knowing smile.

"Just remember that graduate school is a marathon," he says. "Not a sprint." He slides the stack of papers off his desk.

"I often give pop quizzes in my class, which are designed to make sure that you keep up with the homework." My stomach drops a few levels.

"This is the first pop quiz"—it drops a few more—"but there's no physics in it; just a questionnaire asking you to share a bit about yourself so I can get to know you."

I glance down at the paper in front of me. The first question reads, *What is your favorite coding language?*

A pounding builds in my temple.

Coding language? I haven't coded a single thing in my life. Hand shaking, I write the first language that comes to mind, *Python*, even though I have no idea how it works.

Second question: *Approximately how many lines of code have you written?*

Fraud. Idiot. Tears burn my eyes—I am too stupid to be here, and deep down I knew it; I knew that getting into graduate school was a farce, that I don't deserve to be here.

Around me, the rest of the students wrap up their questionnaires and chat amongst themselves. I scribble randomly on the rest of the questions and blankly get up to follow suit.

"Good," he says. "Now we can get into the fun physics stuff. Let's introduce the concept of specific intensity and flux."

He picks up a black dry-erase marker and draws a diagram on the whiteboard: a large square, with arrows pointing in on one side and out on the other.

"With those at your table, I would like you to try to derive the equation for specific intensity using your knowledge of luminosity and flux."

It has been a full year since I have taken a physics class, and I hardly remember any equations, let alone how to extrapolate from one to another. My research skills are useless when solving contrived, pen-and-paper problems in class.

Turning to one another, the other graduate students begin chattering, scribbling equations on pieces of paper and nodding in encouragement. Silently, I look on, hoping against hope that I blend into the background, that nobody bothers to ask me a question.

Minutes feel like hours, tipping from one to the next, until the pealing of the clock tower finally signals the end of the period. From the classroom, I bolt to an empty park bench under a redwood's shade, away from the wandering gazes of the other students.

Thoughts of inadequacy swirl in my mind. Not only must I bear the weight of my dad's diagnosis and my own, not only do I dream of death and disease—now I have moved across the country to pursue a PhD at one of the top universities in the world, where I am inadequate, ill-equipped.

I don't know how to navigate all of this alone, to juggle it all and paste a smile on my face. Bone-deep exhaustion sinks into me; I am so, so tired.

And so, I decide, I will simply detach.

I go numb.

The throbbing in my chest recedes, the tears in my eyes fade, the goosebumps on my flesh melt, until there is nothing left, nothing to feel, save for the knowledge that I'm alive.

I feel nothing.

I am nothing.

Months pass. A heat wave squeezes the Bay Area, wringing it dry with sweat and drought, and temperatures skyrocket into the low hundreds, forcing us all to shutter our windows and cloister inside.

Everyone says that thunderstorms steer clear of the bay, and so when the first fall storm hits and rain shatters a great pine flanking our house, it takes a whole month before someone from the city arrives to mend the broken branches with an enormous white sling.

Mind blank, I barely register the passing seasons—fall tips to winter, and spring nips at its heels.

My dad tells me that Mexico is wonderful. For the first time in his life, he is practicing yoga, and already he says that his body is changing. The physical strength that chemotherapy eviscerated is slowly coming back, and I can hear how much stronger he sounds when he laughs. He takes monthly blood tests, all of which come back stable—normal. No cancer progression. After each one, I heave a sigh of relief.

When we talk, he often asks if I have begun the monitoring protocol for BRCA, and each time I tell him that I'm not ready. Worry seeps into his voice the more months that go by, but I am paralyzed by fear; I don't want my own oncologist, not when I am healthy and young and free. He sends me articles and radio recordings, and they remain unopened in our text messages.

Clear blue skies become weltering and thick, coating the Earth in a heavy fog. Day after day, I bike to and from class, mist soaking me all the way through. The blanket of icy fog blocks even the stars.

Problem set after problem set, an unending stream of physics that I slog through, boxing my answers in tiny, neat squares that I never want to see again. Sixteen weeks of my course on radiation becomes sixteen weeks of fluid dynamics, which becomes a new semester of high-energy astrophysics and then an entirely new academic year.

I spend most of my free time outdoors hiking and backpacking, if only because out in the mountains, I am no longer myself; I am simply another blip in the ancient landscape of granite and jade.

Exploring the peaks and gullies of California feeds a fire of invincibility that courses through my veins. I cannot get enough of it—the towering mountain spires, the near-death experiences, the stink and sweat and dirt—I am recklessly addicted.

It feels good to feel so small.

Out there, I'm alive.

No specter of illness hangs above me—I am absolutely, wildly free.

I develop an affinity for whiskey and wine. During the weeknights, when I am chained to my desk, toiling away on problem sets, I drink a glass or two, enough to glaze my mind and keep the nightmares at bay.

And the dissociation reaps dividends—one night in January, during my second year of graduate school, my mom arrives on my doorstep and says something that would previously have devastated me.

"I'm asking him for a divorce," she says to me, eyes wild.

As she says it, I feel numb, so numb that I have to pinch myself to remind myself that she is here in California, right in front of me, crying.

Divorce. After thirty years of a marriage of vicious arguments, that painful trial separation while I was in high school, a move to *Mexico*, terminal motherfucking cancer, they are finally throwing in the towel.

And so, when she tells me that she is leaving him in Mexico and returning to Texas, I have nothing left to muster but cold acceptance.

"Okay." I shrug.

That night, I get so rip-roaring drunk that I vomit through the early hours of the morning. In between violent heaves, I rest my forehead against the cool sides of the toilet and allow myself to drift, half-asleep. It is only when I wake the next morning, still gripping the base of the toilet, that I feel the ghost of a touch—my mom, who gently collected my hair and rubbed my back while I was sick.

It's difficult to feel much when you're numb to the world.

I begin to see a therapist, Kyoko, weekly. During one of our sessions, I lean back into the plush sky-blue cushions and observe that I feel like a child playing in a pond, recklessly oblivious about who I splash, what I ruin.

Kyoko tells me that I'm dissociating.

I tell her that I don't care.

And then—a shift.

Taylor, my steadfast, unfaltering friend, moves to the Bay Area.

At the beginning, we're nothing more than friends, two adventurists scaling mountains and giggling together under the stars.

Soon, our relationship—he—becomes something more, and despite my best efforts, I begin to wake up. Not immediately, and not all at once, but slowly, over the span of several months. As the months melt into one another, I share with him my history with Ahmad. Why I sometimes leave on the lights when we go to sleep, and dissociate when we argue, and cannot bear to hear raised voices. With every detail, he listens silently without judgment, rubbing my back and holding my hand, saying that nobody deserves to be treated like that—*nobody*. After nine months, I realize, I am wide-awake and we are deeply in love.

When I tell him about the divorce, we are hiking together through the cliffs in Marin. Fog cloaks the trail in a chill, heavy cloud. Comet has run up ahead of us, and all but the black of his tail disappears into the pale, low-angled light.

"It's been a long time coming," I admit, "but I'm still sad about it."

To our left, water courses down the ravine. Comet reaches his paw tentatively toward the seething current but immediately skitters back. Taylor tosses a stone into the water, and I flinch as it flies by—since Ahmad, I am tremendously sensitive when anything gets thrown.

"Of course," he says, calling Comet over to pet him. "That's normal."

In the strange, alien light, I watch Comet curl into Taylor's legs, and Taylor pause to stroke his ears. In my eyes, tears gather—this budding relationship, my new family, feels like home.

"I don't know how to do all of this alone," I say, wiping my face with my sleeve. "I wish I had a sibling or *someone* to go through this with me."

Taylor looks up from stroking Comet and meets my eyes.

"You're not alone," he says. "You have us."

I hiccup, stride over to them, and wrap my arms around Comet's neck. In answer, he curls into me.

"That's true."

Tenderly, Taylor reaches over and grips my wrist, pulling me into a hug. His icy-green Patagonia is soft when it touches my skin.

"I know your parents aren't my own," he says into my neck, "but I'm here for you all the same."

"Do you think," I whisper, confessing a thorny fear embedded in my mind, "that you wouldn't want to be here if I wasn't as accomplished?"

He pulls away and huffs a disbelieving laugh. "Are you kidding? None of the surface-level stuff matters to me. I'm proud of you, but I love you for who you are, not what you achieve."

I chew on the inside of my cheek, reveling in the jolts of amazement flashing through me: I did not know that unconditional love, one not based on achievement, was possible.

I think back to my parents' many arguments, my mom's screaming and my dad's quiet resignation. Nothing about my relationship with Taylor resembles the broken one between my parents.

As the three of us continue down the trail and disappear into the cloud of fog, a sense of security settles into me.

Deep down, I know that I can face whatever life throws my way with these two by my side.

It is Taylor who finally convinces me to get an MRI for my breasts.

"I love and support you," he says, "but like the oncologist said—it's time."

To my parents' immense relief, on a Thursday morning in the middle of April, when vanishing shafts of sunlight chase us across the Bay Bridge, we drive to UCSF.

Taylor stays in the waiting room while I am brought back to change into a soft-pink hospital gown. I walk in lockstep with a nurse who shows me to Radiology. The echoing white of the walls hurts my eyes.

When we reach a heavy, lead-lined door with a yellow Caution sign, my heart skips a beat. Trembling, I follow her through the swinging door and blink back the biting white light of overhead fluorescents.

"This way," she says, and points to the center of the room, where an enormous magnet encircles a long, narrow patient table.

I climb into the yawning mouth of the machine and lie face-down. The padded trough holding my breasts is cold against my skin. The nurse walks to my side, swabs my arm, slides in an IV, and hands me a pair of plush headphones.

"For the noise," she says.

I try to lie as stiffly as possible, shutting my eyes as the dark gadolinium dye courses through my veins, cold as liquid ice. The noise of the magnet builds, whirring and clicking as the magnet switches from off to on.

For thirty minutes, I try to maintain my breathing. Inhale, exhale, over and over. I imagine that I'm one of those mountains I'd hiked, silent and immovable. By the end, my mouth tastes of dry chalk.

When the nurse escorts me back, she says that I should get my results in three to four days.

"Depends how busy we are," she says flatly.

Finally, I reach Taylor, thrust my arms around him, and bury my face into the nape of his neck.

"Let's go home," he says.

Not an hour later, I'm resting on our couch, massaging my sore parts, when my phone rings.

"Hi Sarafina, this is Dr. Dhawan with UCSF Oncology." That molten weight inside me flares.

"The radiologist just called me and would like you to come in as soon as possible for a biopsy." Every cell inside my body quivers in dread, and then terror.

"They . . . they found something."

CHAPTER XI

UNBREAKABLE THINGS

Some objects in the universe are unbreakable. They're fundamental to our universe, and no matter how much pressure they endure, no matter how we slice and dice, excise, heat, or freeze them, they remain, steadfast in the face of whatever comes their way. They form the foundations of stars and provide the framework for rules governing the cosmos, indestructible and everlasting.

No matter what.

The tiniest building blocks of the universe are irreducible and indestructible. Each one has no internal structure, nothing inside to break down into—they're fundamental in the sense that they are *unsplittable*, the essential ingredients for anything to exist in the cosmos.

Elementary particles, like bosons and fermions, are these building blocks. In general, bosons are categorized as "force" particles, responsible for ferrying forces, while fermions are "matter" particles, responsible for giving objects structure and mass.

Quarks, a type of fermion, come in six flavors, with arbitrary yet fantastical names: up, down, strange, charm, top, and bottom. Leptons, another type of fermion, do too: electron, muon, tauon, and their corresponding neutrinos—electron neutrino, muon neutrino, and tau neutrino. Gluons, a fundamental sort of cosmic glue, bind quarks together to form familiar particles like protons and neutrons. Every quark and lepton has a corresponding antiparticle—identical in every way save for having properties of equal magnitude and opposite sign, like electric charge—and their symmetry is responsible for the proliferation of matter throughout the universe.

When stars die and their light is extinguished, they can either form black holes or degenerate stars: white dwarfs and neutron stars. These degenerate stars are extraordinarily dense, compact objects that, while not unbreakable, are composed of some of the densest latticework in the universe. In many ways, they are like gigantic atomic nuclei, primarily composed of electrons, protons, and neutrons. Electron and neutron degeneracy pressure, respectively, hold white dwarfs and neutron stars together, the only pressure preventing the stars from collapsing entirely.

These degenerate stellar relics are characterized by exotic, wildly strange structures. Neutron stars are theorized to have a crust of crushed atomic nuclei coating a sea of flowing electrons. Just below that, the interior structure is thought to be a "nuclear pasta," the strongest material in the universe. In this pasta, the immense pressures of neutron stars allow nuclei to retain more neutrons than usual, which soon decay into free neutrons that form the neutron degeneracy pressure knitting the star together. At the core, the pressure is so unyielding that strange, exotic forms of matter might abound, like a plasma of quarks and gluons, or even degenerate strange matter like strange, up, and down quarks.

White dwarfs, on the other hand, are theorized to have a crystal latticework structure made of carbon and oxygen nuclei. They, too, have a sea of degenerate electrons below the crust, and crystallize as they cool upon formation. While white dwarfs can ultimately explode, it is the particles within them that remain fundamental and unbreakable.

And then there are the four fundamental forces, the laws of the universe. Gravity, the force governing any particle with mass, acts on large distances; electromagnetism, the force responsible for particles with charge and magnetism, operates on both large and small scales, but is typically only felt on the atomic scale. The strong nuclear force acts between quarks, and the weak nuclear force operates between neutrinos, electrons, and radioactive decay.

These particles and forces make up the Standard Model of Physics, our provisional theory explaining the fundamental aspects of the cosmos. Although we don't think the Standard Model is complete—for example, it can't yet explain spacetime, gravity, or dark matter—so far, the particles that it does predict hold true.

Ideally, we will one day have a Theory of Everything that unifies spacetime, fundamental forces, and elementary particles.

But even then, elementary forces and particles will be fundamental. No matter what the universe throws their way, even if stars violently explode or are ripped apart, shell by shell, falling into black holes, they will remain.

The metallic tang of blood coats my mouth—I've bitten my tongue so hard that it's punctured.

Dr. Dhawan is still speaking on the other end of the phone, saying things about how they classify tumors, but my mind is all static and lurching emotion. My feet carry me out the door, down the steps of our apartment complex, and onto the sidewalk outside.

The sensation of falling sinks into me, and the air whipping across my face, carrying tendrils of my hair into the sky, creates an invisible cocoon around my body. Here, nothing can reach me. Not the voice of Dr. Dhawan, not the unreality of my test results. They exist in some other multiverse, one to which I've shut the door and escaped from, just for this moment.

Finally, words that make sense reach my ears. B1 is normal tissue, she is saying, B5 is malignant, B3 is what mine is, somewhere in the middle: *benign with uncertain biological potential.*

Way up above me, the weltering sky darkens. The air grows laden and heavy, swollen with the promise of a spring storm. Three houses down from me, a child on a bike gazes up to the clouds and the wheeling flock of birds flapping toward their nest. Her eyes widen with fear, and she rolls quickly down the sidewalk, all the way back home.

Still on the ground, I glance down at my white V-neck and pink sports bra and try to register what I'm looking at. Are there tumors sprouting underneath my skin? Spiky, poisonous ends unspooling across my chest?

My risk profile, the doctor is saying, makes her concerned, makes her want to be aggressively cautious.

The sounds of children yelling float through the background of the phone. She tells me that she's picking up her three-year-old from daycare but that she wanted to call me before the end of the day to tell me the news.

"Are you there?" she asks.

I blink once, twice, realizing that I need to force my mouth to work.

"Yes," I say quietly.

"I've gotten them to fit you in for a biopsy tomorrow at ten A.M. Can you make that work?"

Thank God it's soon, thank God I won't have to wait with this paralyzing fear for much longer—

"Definitely." The strength of my voice sounds strange when, inside, I am fracturing into one million pieces.

"Great. I'll call you the second I get the results," she says. "Try not to stress too much."

"Sure."

We hang up, and my promise hangs in the air, utterly absurd. How can I not obsess over what might happen?

Already I am mapping out every single outcome in my mind, anxiety compelling me to consider each possible decision; I want to calculate and assign a likelihood to each one. The possibilities are endless and suffocating—the spot they've found must be cancerous; how could I have waited this long to get an MRI; what if I have cancer at the same time as my dad; what if the cancer has already metastasized; how will I be able to afford my treatment?

A vision of my parents standing atop a grassy knoll, abject misery disfiguring their faces as they gaze down at my headstone, swims in front of me.

My heart beats in tandem with the refrain in my mind—

I do not want to die.

I do not want to die.

I do not want to die.

Minutes later, the birds have all disappeared and the child has retreated indoors, and the sky finally breaks open. A thick fog jackets the unrelenting rain falling from above, and for a moment I simply stand there on the sidewalk, letting myself soak.

Without knowing it, my feet finally carry me up the double staircase, through the entryway, and down the hallway. The door to my apartment is still propped open, light spilling out into the corridor, and when I pass through the entrance, Taylor is waiting for me on the couch.

He meets my eyes, and without saying anything, he knows. In one long step, he strides over to me and wraps me up in his arms.

"I have to go in for a biopsy tomorrow," I whisper.

"We," he corrects me, and squeezes me tighter. Through the terror I feel the warmth of his palms, the light pressure of his fingers, and something in his touch soothes me.

"We." I look up at him, blinking back the tears coursing down my face. Later, I say to myself, I will tell my parents. Once I calm down, when I have enough emotional bandwidth to manage their fears and anxieties.

Still, in a hidden corner of my mind, safe from the raw terror rocketing through the rest of my body, the meaning of Taylor's words settles me just a little bit: I'm not alone.

I'm familiar with the hospital now. When we walk into the waiting room the next morning, that signature bitter scent of bleach and blood hits my nostrils, and I have to force myself not to flinch. Taylor's hand tightens around mine.

I change into the pink papery gown quickly. This time, I remember to remove my lapis earrings and the Eye of Horus permanently dangling from my neck—an Egyptian tradition for protection. Gently, I place them atop my sweater and lock them away.

The shaking begins when I sit in the pale-blue phlebotomy chair. Gripping the armrests harder doesn't help—I cannot stop trembling.

From the other side of the room, the nurse gathering materials for my IV looks at me with concern and mutters something about tea before placing the syringe down and bustling out of the room.

When she leaves and I'm alone, the oppressive scent of the room becomes unbearable—too bitter, too metallic—and I begin to heave panicked breaths. The fluorescent lights overhead beat down unforgivingly and my skin tingles, then erupts in goose-bumps.

I am burning, on fire.

I clench my eyes shut.

"Try this," a voice says, and my eyes open to the nurse in front of me with a blanket and paper cup of tea. Hand shaking, I reach to grasp it and slop hot water on myself.

"It's completely natural to be nervous," she says.

Nodding, I try for another sip and let the heat spool across my tongue. She tucks the blanket around me, and to my surprise it's already heated. The warmth sinks into me, and I heave another, longer breath this time.

She grabs a chair and drags it over to sit down next to me.

"Can I hold your hand?" When I nod, heart in my throat, I feel her warm grip wrapping around mine, anchoring me.

For ten minutes, we sit there motionless together. Then, when my shaking finally abates and she successfully inserts the IV, we walk hand in hand to that thick, lead-lined door and enter Radiology.

This MRI is so much worse. Five nurses are here now instead of two, and they pump in so much dye that I feel the cold leaking across my breasts and through my chest. When I breathe, shards of ice careen through my veins.

After a few minutes, the radiologist says they see the lesion, and one nurse continues holding my hand while another sticks a needle into my breast and pushes local anesthesia to the area. I bare my teeth against the pinching pain and let a few tears fall before clenching my eyes shut.

"Can you see the spot? Does it look bad?" I ask, almost pleading.

"Unfortunately, it's really hard to tell," the nurse says, still holding my hand. "We're about to take the biopsy, so just keep focusing on taking deep breaths. We have five cores to take, and then we'll be finished."

I shut my eyes so hard that pinpricks of black dance across my eyelids. All of a sudden, a scraping sensation descends across my left breast and an intense pressure pushes against me. I suck in a breath.

"You're doing great," the nurse says soothingly. Each time they stick the needle in to collect breast tissue, my breath catches.

"Now we're going to insert a clip that will show us exactly where the region of concern is in the future."

"Okay," I bite out. Her hand tightens around mine as that scraping sensation descends upon me once more, and I'm on the verge of crying out when, all of a sudden, it disappears.

"All done," she says, and pushes me up.

"Did you see anything?" I find myself asking once more.

She smiles and gently shakes her head.

"We'll have to wait for pathology to review the tissue."

I nod, numb. Two nurses work together to clean and cover the area with a sterile dressing, and then they walk me back out and tell me to wait three to five days for results.

"Just try to relax while you wait," the nurse says to me and Taylor in the waiting room. "Worrying about it won't help you."

Taylor says something in agreement, and I lean into him, allowing the buzzing in my ears to finally crest and mute the rest of the world.

When we reach the car, Taylor gently pulls the seatbelt across my chest to avoid the tender parts. Helplessness, true helplessness, winds through me, viselike and unrelenting, and I begin to cry once more.

Taylor's hand wraps around mine, and together we ride in silence all the way home.

One day passes. I cannot move from our living-room couch, my legs leaden. Taylor brings me breakfast, then lunch, then dinner, and when I thank him, my voice is hoarse from lack of use.

The next day is more of the same. I tell my parents, and to my great surprise, they aren't nearly as anxious as I am.

"You're a healthy twenty-three-year-old, little one. Your odds of cancer are pretty low right now," my dad says with a short laugh. He's right—BRCA-induced risk exponentially increases with time, and twenty-three years old is young enough to logically not need to worry too much yet, but *still*—what if?

"But what if, worst-case scenario, the biopsy comes back positive for cancer?" He's silent for a moment.

"It's possible," he says quietly. He keeps talking, but the rest of his words float through the ether, past my ears. I've heard what I needed to hear.

On the third day, I force myself to stand up, walk to my car, and drive to therapy. It's the most I've moved in three days, and an achy tingling is shooting through my legs by the time I climb the steps to Kyoko's office.

When I sit down, I am already crying. In between tears, I relay the events of the last few days, stopping and starting every so often to wipe my nose.

Visions of the future wheel across my mind, and I share them out loud as they form: picking up the phone to be told that I have cancer; concoctions of toxic drugs I'm to be on for countless years; my dad's paper-white hands soothing me as I vomit from chemo—

"Do you think that your anxiety can predict the future?" Kyoko interrupts.

"Well, no," I admit. "But it can prepare me for whatever's going to happen."

"Sure." She nods. "That's one way of looking at it. It can make you study more, or train harder."

"Exactly." I think back to the countless tests for which I over-studied, the fact that I made it to UC Berkeley despite the odds, to get my PhD in *astrophysics* of all things—a dream that would have remained a figment of my imagination had I not put in the anxiety-driven work.

"And that's undoubtedly served you well in a lot of ways, so I can understand why letting that go might be hard. On the other hand," she continues, "torturing yourself now won't change the future." She tilts her head to the side, letting the sheets of long black hair fall down one shoulder.

At that, I blink, taken aback. Her logic contradicts my every lived moment, every time I've subconsciously guessed at the future to gain an ounce of comfort for the present.

But . . . she's right.

Obsessing—catastrophizing—about the future doesn't change the outcome. Dragging myself through the specter of potential pain just makes me live through it twice: once now, and once again if it eventually happens.

"Either way," she says, "you're enduring a lot of pain."

For the umpteenth time that day I begin to cry, but these tears feel different—as though they're a sort of release.

"Are you scared about the biopsy?" I ask.

She pauses for a beat, considering.

"I'm not scared for you," she says. "I don't know what will happen, but I know that whatever will happen will happen. Despite how much we worry."

Then she continues, "But I am sad that you're enduring so much pain and fear."

"How do I *not* have this anxiety?"

She smiles gently. "That's the work, isn't it?"

At eight o'clock the next morning, when the last of daybreak melts into the golden glow of the California sun and the fog has long since ebbed away, my phone rings.

Dr. Dhawan.

I rush from the kitchen to the couch, already hyperventilating, and let Taylor tuck me under his arm before putting her on speaker.

"I have good news," she says, and I suck in a breath between my teeth. "Pathology came back all clear!"

Clear.

Clear!

My mind rings with the word, and I shake my head, making sure that it's still there when I am done.

"We'll need to continue to monitor that area," she is saying, "but for now, you're all clear." Taylor heaves a sigh of relief next to me, then grabs me for a bear hug. For the first time in days, I feel a smile spread across my face.

"What does that look like?" I ask, and lightness soars through me, whisking my body way up next to those golden shafts of sunshine.

"Good question, because this is something I wanted to talk with you about. As it stands, you'll need to undergo an MRI every six months, so we can monitor any changes over time."

"Every six months?" I turn to Taylor, concern already etched into his face. He knows what that would cost, how painful the last week has been.

"At the very least," she says, and any relief I just felt is punctured, until she continues.

"Knowing how difficult this last week has been for you, I wanted to bring up another option: a double mastectomy. Did your genetic counselor bring this up with you?"

"She mentioned that there are surgical options," I say, "but I don't really know what that means."

"Many breast-cancer patients get a double mastectomy when they're diagnosed, in tandem with other treatment methods. However, getting a preventative double mastectomy is an option for patients with genetic mutations who haven't yet gotten cancer."

"So, I'd be proactively mitigating my risk?"

"Exactly. And in doing so, you'd reduce your risk of breast cancer from 87 percent to less than 5 percent."

"I'm sorry." I blanch. "Did you just say 5 percent?"

She laughs, and when Taylor turns to look at me, his eyes double in size.

"Yes, five. Many people choose to reconstruct their breasts after the mastectomy, although some opt not to. It purely depends on the person's preference—for a preventative surgery, there's no oncological reason why they couldn't be reconstructed."

I grip Taylor's hand, trying to comprehend. "So, I can get this surgery . . . a mastectomy . . . and it will reduce my risk to nearly zero?"

"Yes. There are complications, of course—things to keep in mind—but that's the brunt of it." On the other end of the phone, I hear another voice, Dr. Dhawan's child, I assume, asking for her mom to grab a strudel from the toaster. For a moment, a tangle of images spins out before me: Taylor and I surrounded by sunshine-yellow kitchen walls, the tang of raspberry jam wafting through the air, a Pop-Tart with steam rising from the icing on a hand-painted dinner plate, and amidst it all a little girl, with bronze skin and curly dark hair in plaits, sitting at the dining table, waiting for her mom and dad to bring her breakfast.

"What sorts of complications?"

"Well," she begins, "the surgery removes otherwise healthy tissue and replaces it with implants. The surgery is intense, and is often multistep, so patients need quite a bit of time to heal and recover. The reconstructed breasts aren't as lifelike as real breasts—although science is getting close! Then, there's the loss in sensation from cutting most of the nerves in the breast area."

My head is swimming, gusts of relief and possibility sweeping over me. Taylor reaches for the mug of coffee with dogs painted all over it and swallows it all in one gulp.

"I would do anything," I say quietly, "*anything* in my power to reduce my risk of breast cancer. I never want to go through this again."

Taylor squeezes my arm and looks into my eyes, face flushed with emotion—I know he feels the same way too.

"I figured." Dr. Dhawan laughs. "If you want to move forward, I'm happy to recommend some surgeons. That said, I'd suggest doing your own research—many surgeons are outside of the UCSF system, and each one has different specializations. It might be worth comparing to see who is the best fit."

When we hang up, Taylor takes my hand and looks at me.

"Whatever you decide, I support you," he says. "But I want you to know that I saw how difficult this week was for you. I never want you to have to go through that again either." Tears spring to my eyes, and I hug him desperately.

"I feel ready." And for the first time this week, perhaps for the first time since my diagnosis, I do.

When I return to the lab the next day, I clean off my entire desk. I stamp the spare pages of formulas scratched on at odd angles into a navy binder, stack my prelim exam flashcards in neat columns and set them on a shelf, gather my broken pen caps and throw them into the trash. I scrub off the ring stains from the decades-old oak and arrange a new stack of empty printer paper neatly in the middle of my desk. On each page, I sketch a chart with four columns: *Doctor, Specialty, Procedure, Notes.*

From my records, I dig out the orange folder from Texas Oncology and flip through until I find the page titled *Recommendations.* On another blank page, I make another chart, this time with five columns: *Doctor, Specialty, Type of Cancer, Action Items, Appointment Booked?*

That one, I begin to fill out immediately, cross-referencing with the Texas Oncology packet. There are so many cancers for which I'm at risk that soon I'm onto the next page: melanoma risk requires a dermatologist once a year, pancreatic needs an endocrinologist, ovarian and breast need an oncological gynecologist; I even need a cancer-trained dietician.

By the time I've determined a game plan for every cancer I'm at risk for, my head is throbbing.

Then I start on the mastectomy sheet.

Before I know it, lunchtime has come and gone. Other students trickle out to go to class, return to swap notebooks, and leave again. Still, I remain at my desk.

This page is more difficult. I begin to scour the internet for different key words: "preventative mastectomy," "double mastectomy," "reconstruction."

But none of the information I'm finding is centralized—it's scattered all over, some on hospital web pages, some on doctors' personal pages, some by the American Cancer Society. Very little is targeted toward preventative patients, those with genetic mutations.

By now, the sun has begun to drop below the horizon, and a darkness is sweeping into the troughs between the Berkeley hills. The other graduate students have long since gone home, and I have not cracked open one single textbook or written a single line of code.

Despite Dr. Dhawan's recommendations and all of my research, I am utterly lost on how to find a surgical team. There is no directory of expert doctors in the Bay Area, no resource designed to compare their expertise to determine who is a better fit. Neither of Dr. Dhawan's suggestions perform cutting-edge procedures; they're still operating with techniques from three decades ago.

By eight o'clock, I scoot to the edge of my chair and let my head fall between my hands. How can this sort of research be this hard? I would trade a month of physics to gain, just for one second, an understanding of how to move forward.

It's now nightfall, and I have written down fifteen surgeons who perform mastectomies. Most require a separate surgeon for the reconstruction process, but I'm lost on how to find their information or determine which type of reconstruction they routinely perform.

If finding a medical team is this difficult for me—someone with advanced research training, no major time constraints, and all the time and energy one could hope for—how do actual cancer patients, especially those with no research background, get through this?

I let my body bend, head falling to my keyboard, and close my eyes. For a half second, I imagine that the four walls of the lab drop away, the ceiling disappears, and now I'm under the familiar stars of the Bay Area. Just above the Golden Gate Bridge hangs a half-crescent moon. There, pulsating in amber at Orion's left shoulder, is Betelgeuse, and directly to its right shines Venus. Below me is the Pacific Ocean, a low fog just barely kissing the seawater swells eddying hundreds of miles away.

The papers on my desk melt away until there is nothing left between me and the sky and the ripples in the ocean, and I'm remembering Grannell's words—*Make good choices.*

And in this moment, I resolve that I will.

That night, I don't sleep.

I eat a snack, bookmark PubMed and the American Cancer Society, create a new spreadsheet. This one is solely devoted to the most up-to-date research on mastectomies. I watch interviews, take notes on medical journal articles, email three cancer centers at hospitals scattered around the United States.

The sun rises and I'm still at my desk, stopping only to relieve myself and text Taylor that I'll be home later that day. When I rub my eyes, I smear black ink on my face, and when my deskmate strolls in at eleven A.M., she shoots me a look of concern.

By early afternoon, I've completed twelve pages of notes. By sunset, fifteen. Eventually, by the eighteenth page, I've finally gained a solid understanding of my options, the surgeons I should meet with, the procedures I'll need, and I allow myself to return home.

The next morning, I begin to make calls.

The first surgeon's office I call tells me that they're booked out for consults for the next six months.

The next surgeon's nurse balks when I share my request.

"You're twenty-three? Without cancer? Why would you need a mastectomy?"

When I try to explain my BRCA risk factor—that I have an 87 percent likelihood of getting breast cancer sometime in my life, and the biopsy they just took—she interrupts.

"But you don't have cancer yet."

For each office I call, I record as many details as possible. Their responses are all the same: I'm too young to consider a mastectomy, I don't have cancer so don't need to remove my breasts. If I end up getting cancer, they say, then they can treat me.

I have a choice, I want to scream. I have the opportunity to save my own life. Why would I wait?

By the fourth office, frustration and exhaustion coil tightly in my chest. By the fifth, I never want to call another doctor's office again.

When the phone rings for the sixth surgeon, Dr. Anne Peled, my exhaustion morphs into rage. Deep in my core, I know that I'm making the right decision for myself, yet nobody is taking me seriously—nobody *believes* me.

By now, their disrespect is so routine that I could scream.

A tangle of memories bursts forth: that astronomer who told me under the stars that astronomy isn't for me; the faculty member at AAS who refused to take any of his female students seriously, who *proudly sexually harassed them*; the physics professor who joked about prostitution to a room full of white boys—and me.

I am so tired.

Furious.

It's all the same, the disdain. The circumstances are different, but the disrespect—that remains the same.

Dr. Peled's receptionist answers the phone on the fifth ring, but to my total shock, she takes down all of my information immediately.

"We can absolutely fit you in by the end of the week. I know that a BRCA diagnosis is stressful—we'll do everything in our power to help you navigate this."

At that, I begin to cry.

Dr. Peled is unlike any doctor I've ever met. She's slightly taller than me, with chestnut hair and wide, hazel eyes, and carries a quiet confidence that puts me at ease the moment she walks into the room.

She spies my stack of notes and immediately sets the tablet preloaded with her own presentation down on the counter at the back of the room.

"Would you mind if I ask you some questions first?" I ask.

"Absolutely not," she says. "I'm so glad you came prepared with so much information."

For forty-five minutes, we talk. I walk her through my medical history, the other surgeons who wouldn't take me seriously. At that, she looks enraged.

"I'm not surprised," she says. "Over 80 percent of surgeons are male."

"You're kidding."

She huffs a laugh. "I wish I was."

We walk through the potential reconstruction options, complications and risks, caretakers I'll need, and the optimal surgery schedule for me—a three-step process, separated by three-month intervals, that includes a breast reduction to preserve my nipples, the double mastectomy and immediate breast reconstruction, and finally fat grafting to get a more lifelike, authentic look.

"Doing your surgeries in steps will also maximize your chances of preserving sensation in your breasts after a mastectomy," she says.

"Wait," I say quietly. "I thought I was going to lose all sensation in my chest."

"Many mastectomy patients do. But me and my husband, who is a nerve and plastic surgeon, are pioneering a new method to preserve the nerves and graft those that are cut to retain as much sensation as possible."

By the time we hit the hour mark, my head is swimming.

"I don't want to take up too much of your time," I say.

"You're not! You're asking great questions, and I want you to feel confident in whatever decision you make."

When it's been one hour and fifteen minutes—the longest doctor's appointment I've ever had—Dr. Peled tells me that she's a breast cancer survivor.

"I know what it's like to get the call that you have cancer," she says quietly. "I've gone through treatment, had to sort out how to live my life after diagnosis. If I'd had the chance to do this proactively to reduce my risk of cancer, I absolutely would."

Almost as an afterthought, she adds, "Knowledge is power."

At the end of the appointment, we hug.

On my way out, I schedule my surgeries—all three of them. And for the first time since my diagnosis, I feel a semblance of control over my fate.

In the bright light of June, one month after my consultation with Dr. Peled, when Berkeley is cast under sunshine and sapphire skies, I begin to train for my surgeries per her recommendations. Three days a week, I'm in the gym lifting weights, and the other four days I'm doing cardio.

My parents are, for the most part, supportive of my decision to get a mastectomy. Still, my mom is stunned that this is the pre-scribed course of action.

"With all of our technology and medical advances," she says over the phone one night, "how is amputation the only way to deal with this?"

For that's what a mastectomy is, she reminds me, fear lacing her words. I try to tell her that I can't shoulder the weight of her anxieties right now, but I'm not sure she hears me.

I'm in the lab every day, trying to fit in as much science as I can, but I'm barely making progress. I'm distracted—in every waking moment, I'm obsessing over my surgeries. When I close my eyes at night, I dream of waking up in a body that's not mine. Some nights, I dream that I never wake.

One cloudless night at the end of the month, a group of grad students goes out to a hip tapas bar in Oakland. Under glittering red lights, they share stories, nodding along faintly to the beat of the bass. One girl just had a terrible Tinder date, and the others are giggling together at the details of her match—paunchy, older than expected, and too lame to take seriously.

With my back facing the street behind me, I watch them, the red flickering across their faces. Each person in the group seems alien, the events taking up space in their minds so unlike my own. It is as though I'm watching them from across the room, like they speak a language I don't, and for a split second I resent them, their happy problems, their ease. It takes me a long time to fall asleep that night, and when I finally do, I dream of my dad.

The next morning, when I'm still curled in bed and Comet is nuzzling into the crook of my knees, I write a long Twitter thread sharing my BRCA diagnosis. At the end, I encourage others to get genetic testing.

"Knowledge is power," I type.

Bile burns my throat, and I press Send. I do the same on Facebook and Instagram, click my phone to Do Not Disturb, thrust it under the pillow, and drag myself to the kitchen.

One scoop of protein, find the knife and chop three strawberries, compost the heads, grab one handful of blueberries, a splash of almond milk, top it all with a tablespoon of chia and flax seeds, close the blender, and the blades whirl. My body moves independently of my mind, calming my nervous system.

The smoothie is chunky and difficult to swallow. I force it down, then trudge to the only desk in the apartment and decide, finally, to check my phone.

One hundred and fifty-seven notifications. Dozens and dozens of direct messages, and they're still coming through.

THANK YOU SO MUCH FOR SHARING YOUR STORY. MY MOM DIED OF BREAST CANCER, AND I'VE BEEN TOO SCARED TO GET GENETIC TESTING. I'M GOING TO MAKE AN APPOINTMENT NOW, THANKS TO YOU.

Another.

I'VE BEEN FEELING A WEIRD LUMP IN MY BREAST, BUT I'VE PUT OFF MAKING AN APPOINTMENT WITH MY OBGYN. GONNA DO IT TODAY. THANKS FOR SHARING YOUR STORY.

Another.

I JUST WANT YOU TO KNOW THAT I THINK YOU'RE INCREDIBLY BRAVE. I DON'T KNOW IF I'D BE ABLE TO MAKE THE SAME DECISION AS YOU.

Another.

Another.

Another.

My eyes burn, and Grannell's words burrow through me for the second time in months. *Make good choices.*

I'm trying, I want to tell her.

August 7, the day of my breast reduction, sneaks up on us.

Before I know it, Taylor is driving us to the hospital across the Bay Bridge at five in the morning. Even then, there's traffic, and the earliest rays of sunlight glint across the cars crowding the eight FasTrak lanes at the edge of Oakland.

Once I check in, they pump IV Ativan into my veins, and I'm so relaxed that the lines of the TV in the waiting room are rippling in front of me.

Dr. Peled walks in at six and makes a beeline for me.

"How are we feeling?" she asks, smiling even at the early hour.

"Great! How are you?" The words fall out of my mouth on their own accord, my brain lagging half a second behind.

"So good. I took my babysitter to see *Hamilton* last night, and it was so fun." She laughs and turns to say something to a nurse before wheeling back around. "Have you seen it?"

"Not yet," I say. "But Taylor really wants to go." I find myself smiling back at her, mirroring her enthusiasm.

"Oh my God"—she stops everything she's doing and makes eye contact with me—"you *must!*"

I find myself giggling, partly thanks to the drugs but mostly at her excitement, and any remaining nerves not already dulled evaporate.

By the time the nurse wheels me into the OR, I'm entirely relaxed knowing that Dr. Peled will be the one operating on me.

I trust her.

It takes five weeks to heal.

For the first week, we watch the bruising on my breasts turn from mottled black to the color of desiccated mold. Taylor writes out my medication schedule on a white-and-black sticky note that we set against a candle on my bedside table.

"It's good practice for the big surgery," he says one morning while gathering my bottles of pills.

Together, we watch so many episodes of *Jane the Virgin* that he begs me to put on anything else. I joke that I'm the one who had surgery, and he can choose what we watch when he decides to go under the knife.

Still on the couch the second week, I decide to make use of my restlessness. I create a centralized resource of my BRCA and breast surgery notes and paste it on my website. Within minutes, a woman emails me that she read about my story on social media, checked her breasts, and was diagnosed with breast cancer. She thanks me for helping to save her life.

Still, not all of the feedback is positive. Several people email me that I'm mutilating my body and sacrificing my worth as a woman. Others suggest that I'm getting a free "boob job." But none of them diminish my resolve—I know what the science is telling us, I can see my risk factors and understand that, most likely, it is not *if* I'll get cancer but *when*, and a few surgeries have the power to alter that potential future forever.

The pain dulls to more manageable, dull throbs. I learn how to tape my breasts around the incisions—one under my breast, one from the base of my breast to my nipple, one around the nipple.

At the three-week mark, my bandages officially come off. Dr. Peled—Anne, she tells me to call her—says they're healing beautifully. After a lukewarm shower, I catch myself looking at them in the mirror. Still swollen, dark-green bruising fading to blotches of yellow and orange. I wonder what they'll look like when they're cut off.

Blend a protein shake, pedal on the elliptical, walk a mile to and from the lab shuttle, type on my keyboard, post on social media. Drive half a dozen times over the Bay Bridge to Anne's office, tape my incisions every morning, rub Arnica gel on the bruising—no, I can't raise my arms over my head; please, would you mind lifting that.

With each passing sundown, time speeds up, until the weeks wholly blend together and two months go by and all of a sudden in five days is my mastectomy.

Fear begins to seep in.

"I can't reconcile the notion of going into the hospital with one body and coming out with another," I confess to a friend.

When I wake up in the mornings and retape my breasts, I touch my skin. Strange to excise healthy flesh, I think to myself. When Taylor catches me crying, he sits down and silently wraps his arms around me. I think of my dad.

"How do people manage this alone?" I ask.

He doesn't answer, but Taylor's grip around me tightens.

The night before my mastectomy, my dad flies into the Oakland Airport.

We drive to a restaurant perched on an overlook in the Berkeley hills. The rapidly dimming sky braces over the gentle slopes, and we stand on the balcony to watch the light sputter out below the horizon, making way for a full moon.

Taylor excuses himself to get a whiskey neat, and my dad sidles over to me.

"How are you feeling about tomorrow, little one?"

Still looking out into the twilight, I say, "I'm ready to get it over with."

"You know," he says, "while I'm so sad you have to go through this, I'm proud of you."

I break my stare into the distance and turn toward him as he wipes his eyes.

"Thanks, Daddy," I say, leaning into him. He pats my arm, and together we look up at the night sky.

"Do you think any of this matters"—I gesture toward my body, then out over the skyline—"in the grand scheme of things?"

"I think it's human nature to seek meaning," he says quietly. His glasses glint in the moonlight.

"That doesn't really answer my question."

"Doesn't it?"

We drive back to the apartment, and my dad bids us goodnight. When Taylor and I finish repacking my hospital tote bag, I sink into the bed pillows and snuggle Comet.

"I don't think I'll be able to sleep," I confess.

"Try," he says. "You're going to need your rest."

In the darkness, after his breathing becomes rhythmic and steady, I pull the sheets up to my chin. From a corner of my mind forms an image of the winding branches that got me to this moment, one trillion limbs over the generations. When I zoom out, I see patterns lighting up like neural networks, one decision leading to the next, and it all seems predetermined, the organization so obvious and clear, until I realize, with a sharp inhale, that what matters is my agency—my choice.

In that instant, time freezes and I see not just the patterns but the randomness, too, the interconnectedness of it all, how we cannot have order without chaos, destiny without agency, fate without choice.

It's the difference, I realize, between walking into tomorrow's surgery lamenting my genes, my fate, and recognizing my ability to choose from the cards thrust upon me.

It's how my dad has maintained his sanity throughout his diagnosis.

It's how I will too.

CHAPTER XII

BIRTH OF THE UNIVERSE

At the dawn of time, the universe was born.

We don't know how the universe came from nothing. Perhaps there was a quantum fluctuation that, over the scale of time, happened to appear long enough to jump-start the birth of the universe. Or perhaps a previous universe collapsed in a Big Crunch, exploding in one enormously vast cosmic firework and kick-starting the birth of another.

But 13.8 billion years later, we know that the universe exists. We see it all around us—in the night sky, in our friends, our forests and deserts and oceans and stars.

The Big Bang is the earliest moment in our universe. It is, in physicist Sean Carroll's words, the moment before which there were no other moments, the moment when space itself—our entire universe—was born. It occurred everywhere all at once, because before that there was nothing at all.

It's extraordinarily difficult to determine what happened in the first minute after the Big Bang, to know anything other than the fact that the universe was hot, dense, and smooth. It's possible that in the earliest fractions of a second after the Big Bang, or a mere fraction of a second afterward, an effervescent spacetime foam consisting of tiny, quantized packets of space and time fluctuated in and out of existence.

Current models suggest that within one-thousandth of a second after the Big Bang, the universe underwent a period of rapid, exponential expansion in a process called inflation, allowing the universe to expand everywhere all at once, dragging any random fluctuations along with it. During this time, it's thought that all four forces of the universe were intertwined in one Grand Unified Force, but they began to untangle as the universe expanded and cooled.

In the first few minutes after the Big Bang, a primordial soup of elementary particles floated around, continuously annihilating with their antiparticle pairs. For reasons yet unknown, the symmetry between particles and their antiparticle pairs was broken, allowing for a slight excess of particles to escape annihilation. These particles are the first building blocks of matter in the universe, forming nuclei and the first ever hydrogen, helium, and lithium atoms in the universe.

Over the next 380,000 years, the universe was opaque, veiled behind a fog of charged particles slamming into one another. But as the universe cooled, electrons and protons were ultimately able to couple with one another in a dance called *recombination*—a misnomer, as they were coupling for the first time—to create neutral hydrogen. To this day, the imprint of recombination continues to shine through a faint microwave signal seen in any and all directions. By the end of recombination, photons emitted from the creation of hydrogen were free to travel through the universe. This afterglow is known as the cosmic microwave background, which has cooled and stretched with time, imprinted on the fabric of spacetime and ever-present throughout the cosmos.

It is possible that this is just one beginning of many. There might be many different universes popping into and out of existence, perhaps with different types of elementary particles or previously unthought-of forces. They might have different measurements of physical constants, like the speed of light. A proton might weigh something entirely different in their universe.

We investigate our own beginnings with the understanding that there might be many other universes, perhaps an infinite number of them running in parallel or occasionally, delicately bumping into one another.

By understanding the beginning of our universe, we uncover fundamental truths about how and why elementary particles and forces are the way they are. We are the way we are because of the universe in which we live. But there is also the potential for an infinite number of new beginnings, unrealized potentials that produce an entirely alternate reality.

All of them beautiful.

Behind me, Taylor and my dad follow into the hospital under a watercolor-streaked sky. My tote bag swings against my legs, carrying my favorite stuffed animal, Stewy Bear; Polaroids of Comet; a navy silk eye mask; my orange Texas Oncology folder of paperwork, now filled with medication instructions; an extra-long phone charger; a compact hairbrush; one of my dad's old cashmere sweaters; and a seatbelt protector.

"Just like waking up at six o'clock for a tennis match, right?"

my dad said when we got into the car, grinning and lovingly jostling my arm. "You ready for warm-up?"

When I rolled my eyes, he just grinned.

I specifically requested that my dad, not my mom, accompany
me for my mastectomy.

"This isn't just my journey—it's Dad's and Grannell's too. This
bond between Dad and me isn't something I chose, but I want
him there all the same," I told her. I didn't say the rest: that right
now, I have no space to manage her moods or juggle her needs.

"But I want to be there," she protested. "I'm your mom."

At this, my heart cracked. Every cell in my body wanted to say
that I wished there was a world in which she could be here too.

But there is no other world, just this one.

The hospital waiting room is empty, save for the receptionist
behind a computer monitor at the desk. She smiles up at me as I
walk up to her.

"Can you tell me your full name and date of birth?"

"Sarafina El-Badry Nance, 3/25/1993." She nods, types away at
her computer, and asks without looking up, "And what are we doing for you today?"

"A bilateral mastectomy," I say confidently.

"Yes, with—"

"With nerve grafting."

"Yes, and—"

"Oh, sorry." I laugh nervously. "And immediate over-the-muscle
reconstruction."

She smiles generously at my nerves.

"Perfect! You're all checked in and good to go. In a few minutes
we'll call you to the back to change and get you started."

We sit on cushioned maroon chairs, and I look jealously at
Taylor as he sips water from our sticker-patterned Nalgene bottle.

"Six more hours, and you'll be able to drink as much water as you want," he says, winking.

"Sara—" A cough. "Sara-feena? Sarafaya-na?" a nurse calls from the door in the corner.

"Sarafina, that's me!" I jump up and follow her to a dressing room, where I change out of my leggings and don a billowing green-and-blue hospital gown.

When I'm done, I follow her to the room in which I'll spend the night. Taylor and my dad are already there, chatting on a recliner and an extra chair the staff brought in. I sit on the edge of the bed, and the nurse fits me with an IV, using three strips of tape to secure the tube to my fingers. I can feel my heart rattling all the way up my neck and into my head. I wonder if the nurse can hear it too.

She takes my vitals and reminds me to remove my gold necklace with the Eye of Horus, which I hand to Taylor.

At that moment, Anne slides open the apricot-colored curtain spotted with snowflakes hanging across the door. In her uniform of navy scrubs, she looks like a commander readying for war.

"We ready to do this?" She grins at me, and I smile back.

"I'm ready," I say. And I am.

She and my dad introduce themselves to each other, and my head swims. Three of my favorite people all in one room. Even though I'm about to lose my breasts, I'm . . . happy.

Countless months, hours in the gym, specialist and therapy appointments, sidelong looks in the mirror and debilitating anxiety—all of it has led to this moment, a moment that I *chose*.

Before Anne and I walk together to the OR, I hug her.

"Thank you," I say.

Because the surgery is so invasive, I'm to receive a thoracic para-
vertebral nerve block to minimize pain before I go under anesthesia. When we get to the OR, Anne shows me how to sit on the
operating table, a nurse spreads local anesthetic, and the anesthe-
siologist picks up a 22-gauge 10-centimeter-long needle and sticks
it into my spine.

"What kind of music do you want to listen to?" Anne asks me.

I think for a moment. "Do you have Harry Potter?"

Ziv, Anne's husband, laughs a warm, startled laugh.

"That has to be a first," he says. "We've got everything, don't
worry."

Anne squeezes my hand and returns to prepping, and the an-
esthesiologist shows me how to lie down comfortably.

In the background of the lilting music is the bustle of the operating room. Three nurses are assisting Anne and Ziv, and they
move around their tools with choreographed precision. The sound
of their well-practiced dance is comforting.

In the moment between light and dark, I feel the phantom touch
of Grannell at my side, the ghost of my dad's hand holding mine,
and even though he is just outside and my grandmother is long
gone, they are here with me, at one of those branches of time
circling in on itself.

Make good choices.

On Ziv's cue, I close my eyes and begin to count down from
ten. As the room darkens to black, I imagine that the operating
table is the soft grass of West Texas, the strange, overhead light
is the moon, and I'm lying under a blanket of stars. The anesthetic gas smells of strawberry, and I breathe it in greedily,

counting the stars backward as I fall asleep: ten, nine, e-e-eight, s-s-s-s-s-seven—

My mouth is bone-dry. I lick the cracks along my lips, then the roof of my mouth. The room is foggy, steeped in haze.

"She's awake! Hey babe," says a floating voice.

From the other corner of the room comes the voice of my dad: "Good morning, little one."

A blurry shape rises from a chair and moves next to me, picking up my glasses and placing them gently on my face.

"Better?" Taylor asks, then sits down to my left and takes my hand in his.

Nodding, I gingerly raise my other arm. My chest is covered in white dressings and tape, and on top of my reconstructed breasts are two pale-blue ice packs.

Ever so gently, I touch my fingertips to my dressings.

Light pressure and sensation spreads all across the upper part of my breasts.

"I can feel them!" I yell. Laughter bubbles out of me, and I immediately grunt in pain.

"Take it easy, little one."

A nurse strides in and glances at my readings. "How are you feeling?"

Suspended in a dreamlike state, I pause.

Is this real?

No more breast biopsies, no more fear of breast cancer or death, the surgery is done and behind me—

"I feel so proud of myself," I blurt out, overcome.

"Yeah?"

"So proud. It's done! I did it!"

Taylor squeezes my hand.

"We did it," I correct myself.

"No, you did it," he says, and slips his fingers through mine once more.

From the back of the room comes my dad's voice. "You did it. You."

I drift in and out of sleep. Every time I wake, Taylor is next to me, holding my hand.

At some point, a nurse brings over a menu and tells me I can order anything off of it. I choose sushi, which makes everyone laugh. I fall asleep trying to stuff a California roll into my mouth.

While helping me to the restroom, a nurse teaches me how to milk my drains. Two of them stick out of me, stitched into my breasts draining blood and pus. Every few hours, I'm to squeeze, measure, and empty their contents. I'm too exhausted and sore to feel squeamish about the maroon fluid seeping out of me.

By six o'clock the next morning, my dad wheels me out of the hospital, and Taylor gently sets me into the front seat. Every time we hit a pothole I whimper, grateful for the cushion wrapped around the seatbelt pressing tightly against my new breasts.

In between hours of drug-induced sleep and bursts of lucidity, I begin to heal. Everything takes place behind a foggy haze: snacks of Goldfish, crumbs splattering across my jade silk robe; count-less episodes of *Love Island*; two heart-shaped elbow pillows elevating my arms and a pregnancy pillow wrapped around my back; Taylor pulling my hair back into a ponytail, rubbing dry shampoo through the ends; pain shooting like lightning through my chest; cups upon cups upon cups of thick red pus draining from my insides.

In seven days, we cross the Bay Bridge once again and Anne removes my drains. Back at home, Taylor cleans my wounds with bacitracin and gauze and goes pale at the gaping holes left over. When his fingers touch a still-numb part of my breast, I see his hands on my body yet feel nothing, and the cognitive dissonance makes me immediately throw up.

Yet the energy to mourn my lost breasts escapes me. Despite the swelling and mottled bruising and still-numb parts, every moment I look down at them, gusts of relief and pride swoop through me.

It's over, I did it, I never have to think about breast cancer ever again.

Ten days in, the dressings come off and when I see my reconstructed breasts for the first time, I cry with relief.

My still-healing scars are a symbol of my bravery. My body, a beacon of strength. My resilience—and yes, my anxiety—got me to this moment.

Funny, I think to myself, that I expected to feel outside of myself upon waking up from surgery in a new body.

As it turns out, I feel more settled in myself now than ever before.

On a Tuesday afternoon two weeks after surgery, the sun slants through the long-paned windows of our studio apartment and glows softly along the planes of my dad's face. From outside comes the chittering of hummingbirds and the whistle of BART rumbling from one side of the bay to the other.

Taylor left with Comet to hike in Marin, taking advantage of January's crisp sunlight and cool breeze, and my dad and I are sprawled across the living-room couch. I'm on my third ice pack of the day, and just swallowed an Ativan to dull the nerve pain shooting across my reconstructed breasts.

Love Island plays in the background, and when I look up from my phone, my dad is crying.

"Dad? What's wrong?"

He removes his glasses and wipes his eyes.

"I'm just really sad"—his voice breaks—"that you have to go through this."

"I know, but it's okay! I'm okay!" I try placating him, fighting against the dizziness fogging my vision as I reach out a hand toward his.

He shakes his head.

"I know you are." He shakes his head again, as though ridding himself of a particularly intrusive thought. "I just wish it were different. I never thought life would be like this, you know," he confesses, and a knife slides into my chest and between my ribs. I fall silent.

"If I'd had a choice, I never would have passed down this mutation to you." His voice is raw. "But I didn't have that choice. The science just wasn't there."

Almost as an afterthought, he adds, "But that's the way life is."

He wipes his face, and I grip a ceramic mug of tea to stop my hands from shaking. Outside, the fog rolls in, and the sun glints through the strange, refracted light. For a moment, he's silent.

"I've had to grow and adjust," he begins, "just as I know you have. One of the things I've realized—and I hope you know that I've worked hard to get to this point—is that cancer is one of the most beautiful gifts that I've ever received."

Despite myself, I frown.

He continues. "Everyone has a different relationship with their diagnosis. Mine won't be the same as yours, and it won't be the same as others affected by cancer." He shrugs.

"But cancer allowed me to return to what's important to me. After decades—a lifetime, really"—he laughs an embarrassed laugh—"of chasing success and achievement, I've realized . . . none of that bullshit matters."

The words he's saying are foreign, strange. Is this the same man who drilled me on the tennis court day after day? Who sat with me at the kitchen table solving math problems deep into the night, who checked over my homework to make sure I made no mistakes? Who encouraged me to go into physics, to stick it out despite it all?

My confusion must have shown on my face, for he laughs again.

"Not the same guy you grew up with, huh?" And for a split second, a small part of me roars inside—where was this man, this compassionate, empathetic man who loves me *no matter what*, when I needed him as a young girl?

As though reading my mind, he smiles sadly.

"I've grown a lot," he says, "and I've made a lot of mistakes. But I've learned a lot too." His breath catches.

"Like what?"

"After I was diagnosed, I spent a long time trying to understand what all of this is for." He gestures around the apartment, points at the television, the uneven stack of physics textbooks, Grannell's acorn-and-eucalyptus dresser, the violet orchid standing proudly in its planter.

"I've tried to understand what makes a good life," he continues. "I realize that for me, it's about appreciating beauty. Living in the present. Learning to be of service, vulnerably and authentically, to others. To you," he says simply.

Through gaps in the thick cactus arms sprouting outside the windows, the fog thickens, casting the earth in a dreamlike haze, and for a moment everything pauses, suspended in time, and the world stops turning. Even the birds cease chattering, tarps stop slapping against the breeze, not one single car honks.

I look into my dad's face, throat tightening, eyes burning.

"In Spanish, there's a word: *fluir*—the flow of the ocean, the flow through life," he says. "I'm learning to live in the moment, to practice *flujo*. Life just *happens*." He scoots toward the edge of the couch and leans toward me.

"Our ego wants to control it, to change it." Staring into my eyes, the blue of his irises is blinding. "But it can't. Nothing can. Every moment, every day, every breath—all of it is a gift. There is no good or bad, it just is.

"I'm proud of the choices you've made, little one. I know it hasn't been easy. Just remember that I love you, and I'm always here."

In a heartbeat his arms are around me, wrapping me up in that familiar scent of pine and oak, and I let the tears roll down my face and land in the knit of his sweater.

"I love you too," I whisper. "I'm still learning."

"That's what life is about!" he says, laughing, and with that sound the world spins on its axis once more; sparrows warble; dogs bark; a bus brakes, expelling a faint hiss. "*Flujo.*"

Three days later, Taylor, my dad, and I stand on the curb of Southwest Terminal 2 under the weltering sky to say goodbye. I watch my dad throw his arms around Taylor and whisper something into his ear, winking. Inside, I feel my heart swell so full I'm sure it will burst.

My dad hugs me delicately, trying to avoid pressing up against any sore parts. A perverse sort of nostalgia blooms in me, and for half a breath I wish for another surgery, another night in the hospital, so that my dad won't leave, can stay by my side.

"Remember," he says into my ear before stepping through the sliding doors. "I'm always here, perched on your shoulder." He taps at my clavicle, as though expecting a miniature ghost to condense with his features, his twinkling gaze.

As he disappears under the dangling blue-and-white Southwest sign like a giant spectacled cheetah sprinting away, he shouts over his shoulder: "Be a student of life, not a student of death. *Flujo!*"

On the drive home, leaves gust in imperfect, eddying circles, speeding up until they're all adrift and the sky breaks open. Fat, cold droplets land on the windshield before wipers whisk them away, and I sink into the leather of the seat, regret and abandonment twisting like vipers in my core.

At first, the anxiety-riddled criticism piercing my mind is familiar: Why didn't I take better advantage of our time together? Why did I waste so much of it watching TV, sleeping, complaining? For God's sake, why didn't I do a better job of comforting him? And the thought above all the others: *Damn you for being so selfish; this one is over, and you now have one less visit together before he dies.*

From the periphery of my vision comes the faint awareness of a warm pressure engulfing my hand, rubbing tenderly along my thumb. Taylor.

"I just—" I try to say, but nothing comes out.

Taylor continues stroking my thumb, and the rhythmic motion slows my thoughts, until I remember my dad's words. *Every moment is a gift.*

I watch the droplets thrumming against the car, transfixed.

Two, four, twelve, I quickly lose count. So small, so many. An infinite number.

"Are you okay?" Taylor asks.

With each drop sliding down the windshield and splashing onto the asphalt, the thought clangs through me that they, too, are transient, countless, their end inevitable.

Through the mist of the storm, the drops glint, halolike, in front of headlights whipping past. In the haze, they look like stars.

"Yeah," I say. "I am."

It takes six weeks for most of the pain in my breasts to disappear. Random nerve zaps linger, zinging through my chest, and an aching throb tugs at my tissue after jogs, but I'm *running*, walking, climbing, reaching!

A fervor engulfs me: up at five forty-five, work out at seven, lab by nine, nights are reserved for my budding social media, interviews with NPR and the BBC. I juggle between astronomy science communication and women's health advocacy, where I share about my gene mutation and surgeries. Both topics take up space in the forefront of my mind—this, I think, is what it means to have purpose. I book a TV show, pull all-nighters in front of a telescope, drag myself to and from my computer, skip meals.

With each published article and completed interview comes the gnawing dread that it's not enough—*I'm not enough,* the ceiling keeps floating higher, and I cannot reach it. All I can do is continue to hop up and push harder and hope that one day I'll achieve enough, but the understanding is beginning to sink in that *there is no enough,* and I am a hamster on a spinning wheel, feet flying, heart racing, going nowhere but in circles, until I am used up, empty, and have nothing left to give.

What am I worth then?

Manic, unable to sit still for more than a half hour, I push harder. Code remains illegible, spectra don't reveal their secrets; still, I ram my head into the wall.

Taylor and I begin to fight.

"You need to rest," he tells me.

"I've been resting for the last six months," I snap back.

"You were recovering from *surgery*," he insists, exasperated. "That's not rest."

Unease trickles through me, descends down my spine. If I don't work so hard that I'm breaking apart at the seams, then who am I?

One afternoon in late February, I trudge up the steps to Kyoko's office. Shafts of sunlight fall across the top stairs, and I pause for a moment, relishing the feeling of sun across my face.

"Come in," she beckons when I reach the top. As I cross the threshold into her office, the calming scent of sandalwood and patchouli embraces me.

She's designed it thoughtfully: two dark-green succulents in textured white planters bank a moss-colored love seat with four pillows piled onto its sides and pushed up against the back wall. Above it hang three framed posters in neutrals and watercolor, to the right are shelves stacked with notebooks, an unsigned postcard, beads of ocher and cobalt, sticks of incense, and twin Japanese figurines.

In front of the sofa is her usual ivory-white armchair, where she perches comfortably, and behind her are three paned glass windows cracked open for airflow. A small humidifier hums in the background.

We exchange the usual pleasantries—*How has your week been? So glad we're finally getting some sunshine, did you hear how much snow Tahoe's been getting?*

"How have things been going this week?"

"I've just been working a lot," I say, twiddling the pillow's ends in my hands, eyes heavy. "I'm exhausted. I have so much to catch up on."

She nods. "How is that work going? Are you enjoying it?"

"For the most part." I pause. "I love the astrophysics—the actual science of it all."

A whirlwind of emotions whips around inside me as I continue.

"Taylor says I need to rest. Part of me *wants* to rest. But I don't know how to *not* work. To take a break. When I wake up, there's this pit of anxiety immediately sinking into me. When I sleep, I have nightmares. And when I'm just sitting on the couch, or taking a bath, I feel utterly useless."

"What would happen if you rested?" Kyoko asks gently.

I pause, thinking.

"I would . . . disappear. I would be a waste. Not worth anything," I whisper, salt and tears gathering in my eyes.

"So, for you, working hard is a means of self-preservation. If you stop producing something of value, you cease to exist," she summarizes.

My vision swims. I see those yellow kitchen walls, my dad and me hunched over the dining table, poring over math problems, stars winking into existence billions of miles above.

I nod. "I guess so. And I feel so guilty because it's like there's this part of me that just *pushes and pushes*, and I don't know how to stop. I know we've talked about how my anxiety isn't always a bad thing, how it helps me achieve things and succeed. But it also makes me feel really shitty when I don't."

Win the tennis match and Mom and Dad will stop fighting, *get an A on the paper* and I'll get a hug, *become a physicist* and they'll love me; just until my inevitable failure, and then I'll be forced to start all over again.

"When you were a child," she says, "your parents prioritized your success. You learned that you had to achieve to be seen and loved."

Tears slide down my face, sticking to my cheeks.

The sky's the limit, I remember my dad saying. Does he know I took the advice literally?

"I feel like this anxiety—working hard at all costs—is just part of who I am."

"In some ways," she says kindly, "it is. But it doesn't begin with you. It's generational."

I blanch, thinking of my mom's anxiety and depression, the fact that she, too, has a PhD. The trauma she endured: the boarded-up windows she hid behind in Cairo to escape falling bombs, trying—and failing—to fit into the United States as a brown woman, the War on Terror, being separated from her family, an ocean away from home . . .

"How is that possible?" I whisper.

"Genes remember," she says. "From one generation to the next, they get passed down to help future children handle those same stressful situations."

"And if that stressor is no longer there . . ." I begin, and she inclines her head toward mine.

"Your body might still feel all of those emotions, could be primed to fight or flee, yes," she finishes.

For a split second, an apparition of my mom hovers before me: curly brown hair, like mine; slanting lips; wide, brown eyes; long, delicate fingers; identical statures of identical height.

Are we so different?

"In 2013," Kyoko is saying, "a group of researchers performed an experiment studying intergenerational trauma using mice. They stressed the first generation of mice, traumatizing them with electric zaps every time they smelled the scent of cherries. Several generations later, the mice pups displayed a sensitivity to the cherry scent, even though they never directly experienced the trauma."

"I didn't realize," I whisper, "that this is all related."

Impossible to feel all of this at once. The room is all color and scent, and with each blink of my eyes, a new emotion crests into me: gusts of relief, the building awareness that this isn't all my fault, that it's *inherited*; a deep, crushing sadness at the pain my mom endured; flashes of anger at this prison of trauma—how can one person bear it all?; the sensation of my heart cracking, this is all too much. Those long, narrow fingers of mine wrap around my Nalgene bottle, knuckles white.

With a tender smile she says, "You're working to break free of a generational cycle. That's hard."

I shut my eyes, biting back the tears, and open them to the shafts of sunlight glinting along the windowsill.

"So," I say, "this intense anxiety, my desire for achievement—it's all related?"

"Yes," she says quietly. "It's about survival. Resiliency. And love."

I close my eyes once more, and for a split second I'm back at the McDonald Observatory, unspooling above me is a velvet canopy of stars that flicker against the night. In the distance, a hawk screams, an earthly sound; inhaling, the air smells of limestone dust and cedar, an earthly scent. The mountains towering into the sky are ancient, and when I remember them, I feel small.

"All of this makes sense," I say, and I'm back on the love seat in

Kyoko's office. "But I love the work that I do. I *love* astronomy. Have since I was a kid."

She laughs, a pealing, beautiful laugh. The sunlight glittering against the sable of her hair glows softly.

"Isn't that the wonderful part?"

"What do you mean?"

"What" she asks, "do you love about astronomy?"

I stumble over the words, trying to find the right way to explain.

"I think that what I love most is the perspective. How small I feel when I look up and remember how little everything matters in the grand scheme of things."

She nods encouragingly, like I'm on the verge of discovering some fundamental truth.

"When I feel all of these emotions inside"—I gesture toward my chest—"all of this anxiety, the one thing that helps is reminding myself that the universe doesn't care if I succeed or fail."

"Interesting," she observes, "that you chose a discipline so intrinsically tied to perspective."

"Yeah," I agree. "It almost feels . . . inevitable."

Her eyes cut toward mine, soft yet piercing.

"Yes," she says, "it does."

The sun begins to drop below the horizon, splattering watercolors across her hair, my face. In the pale, sputtery light, the sensation descends on me that we are close, that after this conversation the world will spin differently on its axis, and I will be pulled along with it.

"It's almost like you're exploring the universe out there," she says, "in order to explore the universe within yourself."

Click, the realization slides into the empty, waiting slot in my mind, and 1 trillion neurons light up, a neural network activated, come alive.

"We are a way for the cosmos to know itself." The sentence comes unbidden to my lips. "Carl Sagan," I say.

Somewhere above us, dangling in the black void of the universe, giant balls of gas and fire are exploding, one every single second, blowing their bits into the crushing vacuum of space, their demise our beginning.

"We are the stuff of stars," I say, "born from supernovae. In trying to understand the universe, I am trying to understand myself. Us. Humans."

Kyoko nods, waiting.

"And vice versa—by trying to work out what's inside of me," I say, "I'm better able to understand what's out there."

Some great force sets the planet spinning one degree off-axis, and everything looks different from how it did before; the room is resplendent in lavender, beside me is the phantasm of my mom, and from nowhere at all comes the faint scent of pine and cedar—my dad.

As if in answer, a star flickers into existence.

"The beauty," Kyoko finishes my thought, "is when you can do both."

EPILOGUE

DREAMS

It is profoundly human to explore the unknown. From sailing the tempests of the ocean to summiting the tallest mountains of Earth, human ingenuity has propelled us forward over the millennia, uncovering secrets of our universe that we did not even have the imagination to dream.

Just in the last century, we've invented aircraft to carry us into the skies and rockets to launch us to the moon, engineered a low-earth-orbit space station, and sent robots to Mars. We are, at our core, explorers, seeking answers to the most fundamental questions of humankind amongst the stars. Questions that cast our entire existence into perspective.

Questions like "Are we alone?"

Life on Earth is plentiful. Deep within subterranean cracks beneath the surface of the Earth and in ice crystals dangling from the roofs of caves, even in gold mines and oil fields and vents hissing from the ocean floor, life is positively teeming on our planet. Wherever there is water, even just a drop, there seems to be life.

Are we unique? Are there other life forms out there amongst the stars also seeking contact, waiting to be discovered?

We've launched probes past the boundary of the solar system and out into interstellar space to search for signs of life. In addition to science instruments, two of them, the *Voyager* spacecraft, hold the Golden Record—twelve-inch gold-plated phonograph records, upon which are descriptions of humankind: photographs, songs, speeches, science, even brain waves. The spacecraft are still out there, floating amongst the stars, awaiting discovery by another life form to share our existence with the universe.

It is unclear how likely the existence of extraterrestrial life forms is. The Drake equation, primarily developed by Frank Drake, seeks to answer this question by weighing several factors: the rate at which stars form, the fraction of stars that might be suitable to sustain life, the fraction of those stars that have planetary systems, the number of habitable planets per solar system (those that can retain liquid water), the fraction of these planets on which life could arise, the fraction of life that is intelligent enough to engineer communications with other space-faring civilizations, and the average total lifetime of a civilization.

But even with an equation, the data inputted into the formula is never precise, and thus varying answers arise, from exceedingly unlikely to predicting that there must be millions of planets that have the potential to harbor life.

Bodies in our own solar system hold the potential for extraterrestrial life. Mars is an especially exciting candidate, given that water ice abounds on the surface and beneath polar ice caps. Enceladus, a moon of Saturn, has water jets and a subsurface ocean, and Jupiter's moon Europa, with its plumes of water vapor, icy surface, and potentially even subsurface ocean, might also be strong candidates.

In fact, it appears that water, a necessary ingredient for life, might not be rare at all in the cosmos.

It's possible that we live in a universe primed for life. The elementary particles, forces, and constants (for example, the speed of light) that we understand to be foundational to our universe might be the only configurations that could form life.

But even if there is other life out there, our planet, with its oceans and rain forests, rainstorms and rolling plains, and barren deserts and ice caps, is definitely unique.

There is no other Earth.

No other us.

These thoughts moor us as we set sail into the stars. Already, humans are developing plans to return to the moon and make the leap to Mars. The cosmic frontier is just beginning.

As we explore the stars, plunging into the unknown, we are not simply learning more about our universe. We are discovering ourselves.

We are learning what it means to be alive.

What it means to be human.

Dust particles the color of rust float up, mote by mote, into the atmosphere when I step out of the airlock. Two crew members flank behind me, and together we await permission to begin our trek toward Waypoint 1.

My phone, rewired for radio comms, buzzes. On my right, Bioengineering Officer Averesch clumsily palms his stylus, the EVA suit constricting his motions. Those of us conducting the EVA collectively ignore typos or proper capitalization—it's too hard to type in these gloves.

DELTA: AIRLOCK SECURE.
HABCOM: COPY DELTA.

From the depths of the front pocket on my EVA suit, I fish out my own stylus, cursing at the thick gloves covering my fingers, protecting my skin.

ALPHA: ENROUTE TO CARADHRAS. WILL B OUT OF COMMS FOR 1 HR.
HABCOM: COPY DELTA. BE SAFE OUT THERE.

Thanks to oppressively high winds and our thick, impenetrable visors—excellent for keeping us safe, terrible for communication—it's impossible to hear one another. We've adapted scuba hand signals for EVAs: a rotating flat palm indicates there's a problem; a palm waving downward signals "slow down"; a flattened palm held up is "stop." Immediately.

I flash the "OK" signal to Officer Averesch and Officer Phillips—my thumb and index finger in a loop, remaining fingers extended. In a single-file line, we trudge up the nearest ridge, careful not to slip on scattered shards of loose rock. In the distance, winds begin to whip around in tight, violent circles.

A dust storm.

We'll have to hurry, I think to myself, and I know the other two officers are mirroring my thoughts. Move quickly enough to capture as much science as we can but move gingerly enough to avoid injuries from the lava.

Delicately, I step forward, leading our line, trying to avoid kicking up too much powder and making visibility even worse.

Below us, hundreds of miles of volcanic tunnels wind through the planet's crust. As we summit the ridge and gaze down toward our next obstacle, a sea of shards of volcanic rock, I point two of my fingers toward my eyes, then at the rocks.

Careful.

The last thing any of us want is to fall off the aʻa lava, the spiny, rough lava fragments biting into the sky.

Or crack the surface and fall into one of the many lava tubes.

Through the clear plastic of my helmet's visor, nothing but miles and miles of jagged slate and amber volcanic rock stretch in front of me. Behind us, to the east, is the Hab—home. Follow the ridge as far north as possible, and we'll land at Olympus Mons. South will lead to the smaller volcano, Arsia Mons.

The faint hiss of clean air into my helmet reminds me to wet my already dry lips. No water until we get back to the Hab and can safely flip up our visors. Frowning, I lick my lips again.

Looking down, I pull up the GPS coordinates preprogrammed into our phones and scan the ridge for final confirmation.

Another thumbs-up. We're heading in the right direction—to the far end of the Red Planet.

Mars.

Sort of.

I'm halfway through an analog astronaut Mars simulation on Mauna Loa, one of Earth's largest volcanoes, which towers atop the Big Island of Hawaiʻi. There are five of us making up the *Valoria III* crew, and together we live in the Habitat—Hab, for short—conducting research, maintaining the facility, and surviving "on Mars."

And for all intents and purposes, it's close. At an altitude of 2,500 meters, we're breathing limited oxygen in one of the most isolated, harsh environments in the world.

Out here, treading the volcanic surface on a research mission EVA (extravehicular activity, or spacewalk), it's easy to fool my senses into thinking that I'm actually on Mars. For ten days, I haven't felt sunlight, thanks to the sealed-off Hab and thick, full-body EVA suit, helmet, boots, and gloves. The suit is constricting, itchy. I miss the feeling of sunlight dancing across my skin.

Three of us are approved for EVAs at a time, while the other two crew members remain at the Hab, acting as HABCOM: monitoring our progress through mutually agreed GPS locations and messages. Today, they're worried, too, about the weather.

"Dust storms" are what we call rainstorms on simulated Mars. And even though we're in a simulation, the storm can cause very real damage to the electronics in our EVA suits. Worse, it could separate us from the Hab, leaving us at risk of getting lost or injured.

HABCOM: INCOMING DUST STORM 20 MI. AWAY. MOVING FAST.

Officer Phillips glances up from her phone, which is attached to her wrist, and taps at the screen. I look down, motion "OK," and wave Officer Averesch over. From his backpack I hoist out the laser-induced breakdown spectrometer, LIBS for short.

Officer Phillips nods, seeing the exchange, and palms at the phone, typing a response.

CHARLIE: COPY. ALPHA IS GETTING A FEW SAMPLES AND THEN WE'LL RETURN ASAP.

Within seconds, HABCOM responds.

HABCOM: COPY CHARLIE. BE SAFE.

The LIBS is a heavy black instrument shaped like a gun. Inside are three high-resolution spectrometers, two cameras, a battery, and a high-rep-rate Class 3B laser designed to analyze and identify elements. Portable, immediate analysis across a broad wavelength spectrum—ideal for firing at lava samples strewn across the Martian surface to try to determine what they're made of and where they came from.

Kneeling in the rust-colored lava, gingerly leaning into my kneepads to avoid snagging my suit on the jagged bits, I search for a smooth enough patch. There aren't many—the lava is porous and serrated, and I need enough area to evenly press up against the glass wall guarding the LIBS's laser.

The desire to rub my aching eyes overcomes me. Peering out over the volcanic sea, I can tell the dust storm is thickening, the sky a seething, swirling mat of rain and wind. If there were trees on Mars, they would be swaying; if there were birds, they'd be scattering.

But there are no trees, no birds.

Not even here, on simulated Mars, in the desolate far reaches of Mauna Loa.

For a split second, that familiar pull of anxiety washes over me. If the storm hits while we're out here, we'll be completely stuck at the mercy of the weather. Without solid comms, over 300 meters away from the Hab, with equipment that would be irrevocably damaged, we'd risk severe injury . . . or worse.

Uneasy, I stare at the rock in front of me, vision going blurry.

And then, with a deep breath, the hours of training, the years of therapy, slip into place.

Dozens of hours spent scuba diving deep beneath the ocean surface and navigating through that impenetrable void of blackness, learning to master my anxiety through breathwork and sheer willpower.

Inhale for a count of four, hold for four, exhale for four, hold, repeat the cycle.

One moment, one breath at a time. Not useful to think about what might come next, that is out of your control, all you have is this moment, this breath.

Just keep breathing.

Notice the tightness in my chest, the burning in my throat. Feel my suit itch across my skin, taste the stale oxygen in my helmet, listen to the wind gusting across the volcano, smell the salt in my sweat, see the rolling lava plains, the ropy gray pahoehoe and the spiky a'a.

In this moment, I am here.

In this moment, I am safe.

Another cycle of breathwork. With each exhale, my heart rate slows. The fog descending on my vision clears. My palms dry as quickly as they became damp and slick.

Settling back into myself, grateful that just a few seconds have passed and my crew members haven't noticed, I square my shoulders and press the LIBS against the slab of rock once more. This time, my index finger squeezes the trigger, and the laser begins to pulse.

The reading comes back instantaneously: sulfur, silicon, magnesium, and iron. Exactly what I expected.

The thrill of a successful experiment goes through me. By the end of this, I will be one step closer to understanding the formation of our solar system, and the thought streaks through my body, making me giddy with excitement.

I shift my weight to collect two more samples of the slab, and they go quicker. Minutes pass, and when I look up to signal to the crew that I'm done, a cloud of rain—the "dust storm"—grows darker, inching toward us.

To Officer Phillips, I flash the signal that I'm ready to return home, and she types in short bursts with her stylus:

CHARLIE: RETURNING TO HAB AREA. PLS PREP AIRLOCK.
HABCOM: COPY CHARLIE.

I stand up and catch my balance on the a'a. In front of me, Officer Averesch leads us back up the ridge with long, quick steps, and in a single-file line we trace our steps back to the Hab.

If we weren't covered in thick EVA suits, we would feel the air getting damper, heavier, twisting with the incoming storm. By the time we reach the Hab, we're gasping for breath, and the dust storm is just minutes away.

ALPHA: ENTERING AIRLOCK.
HABCOM: COPY ALPHA.

The monitor to the right of the door of the Hab flashes green, and we cautiously enter the airlock, glancing to the barometer on the wall—6 millibars, matching exactly the "Martian" pressure outside. Thank God; if it wasn't, on real Mars, we'd be blown to oblivion due to the pressure imbalance between the inside of the Hab and outside on the surface.

On simulated Mars, we try to replicate real Mars as closely as we can—which includes equalizing pressure in the airlock before entering the Hab.

The three of us squeeze into the chamber, facing the door to the engineering bay, and shoot one more message to HABCOM.

DELTA: AIRLOCK SECURE.
HABCOM: COPY DELTA.

DELTA: RECOMPRESSING AIRLOCK.
HABCOM: COPY DELTA. SEE YOU SOON.

Officer Averesch hits the flashing red button, and from the steel barrier of the Hab comes the faint hiss of airflow, repressurizing the airlock to match the Hab. We wait the usual four minutes, the time that it takes to equalize with the Hab's internal pressure, and when the button finally turns green, we tug our helmets off and wipe the sweat dripping off our faces. Officer Averesch—Nils—heaves open the door to the engineering bay, and gratefully we pile in.

Sweat slicks along the sides of my face, and I desperately pull my hands out of my gloves to wipe it off.

"We really pushed the time limit on that one," Nils says. Officer Phillips—Britaney—laughs.

"Seriously," I say, agreeing, and slip out of my suit to hang it above the industrial vacuums designed to clean the suits.

From behind us, Crew Operations Officer Elisha Jhoti and Commander Michaela Musilova poke their heads into the engineering bay.

"It's already storming outside," Michaela says. "We'll need to hold off on using the vacuums. Britaney, can you put the Hab in low-power mode?"

Britaney sidles over to the back left wall, where wires and plugs tangle over one another, and flips a switch.

The rest of us heave a collective, resigned sigh—low-power mode means no lights, no heat, no charging of devices, and limited cooking ability. The worst dust storms are those that last days, when the Hab's solar panels are unable to hold a charge and the Hab reverts to reserve power. Those Sols—Earth days in reality, but called Sols here in the simulation—are long and dark and dull.

Mostly, those are the Sols when I question if I could really make it on Mars.

Gingerly, I slide the LIBS out of Nils's backpack and place it back into its case, making the mental note to charge it when we get power back, slip into slippers to avoid the cold aluminum floor, and stride through the airlock and into the Hab.

All around me is a circular 1,200-square-foot white dome soaring at the peak of a mountain on Mars. Around the bend is the communal kitchen, full of Costco-size freeze-dried cans of food; a hodgepodge shelf of spices; a small, whistling fridge; a dual-burner stove; a microwave; and an industrial toaster oven that has accumulated so many crumbs and motes of dust that it looks as though it's sat there since the '60s. In front of the kitchen is a long white picnic table, with a sixty-inch screen hanging above displaying the Hab's vital energy consumption and CO_2 levels.

The center of the downstairs area is designed efficiently to accommodate work and exercise. Six modern desks, each equipped with a new monitor, keyboard and mouse pair, and desk lamp, toe the sweeping curve of the Hab. Directly behind them, exercise equipment peeks out from underneath the steel staircase leading upstairs.

Around the bend, squeezed behind the downstairs bathroom, is the lab. As bioengineering officer, Nils spends the most time there, using a bioreactor and genetically engineered microbial cell factory to research bioplastics on Mars.

Up the staircase are six bedrooms, each with a twin-size mattress and miniature desk that are nestled underneath a slanting ceiling and separated from one another with paper-thin plaster.

Home.

We type up our mandatory daily science and mission reports quickly, trying to conserve power. When the comms window opens, I promptly shoot an email to CAPCOM (ground support control center) and also type a quick note to Taylor, my now-fiancé, who proposed to me during a hike just over one month ago.

If I have to eat one more meal of freeze-dried noodles I'm going to scream.

Knowing I won't immediately get a response thanks to the twenty-minute Martian time delay, designed and implemented to re-create the time it takes light to reach Earth from Mars, I shut my laptop and trudge over to the kitchen.

Through the one circular window in the Hab, between the workspace and the kitchen, I pause and see the veined Martian surface glazed with fog. Sheets of lava interweave with one another, forming a latticework of underground tunnels running invisibly beneath the Martian boulders.

Impossible that I am here. Using a laser spectrometer to collect data on supernovae, traversing the Martian surface, eating freeze-dried food with the crew, surviving through a dust storm.

The sky above is an impenetrable, opaque mat of silver. With blackness closing in quickly, for a split second I see myself from above, perched at the window of the Hab, gazing out onto a foreign land, a planet dangling in a sunbeam 120 million miles from home.

Dinner, to my relief, is quinoa, wilted red peppers and spinach, and black beans that took two Sols to cook. When I chew, they're so dry I suppress the urge to spit them back onto my plate.

"The beans are"—Britaney remarks in the fading light, searching for a kind word—"different." Elisha, who cooks meals with me, snorts.

"At least it's not another night of Space Noodles," she says. Nils grunts in agreement.

"I like Space Noodles!" Michaela protests.

I grab the cayenne pepper bottle at the center of the table and unscrew the red lid.

"Space Noodles are bland." I shake the flakes onto my food. "Slimy." Shake. "And revolting," I finish, setting the bottle back on the table.

Elisha raises her eyebrows at me.

"Sorry." I wince as the table erupts with laugher. "It's true!"

When I hand our dishes to Nils and Britaney, the cleaning crew for meals, I spot a streak of dirt along my thigh. The desire to wash myself gnaws at me, and I force myself to ignore it.

The "restocking rocket"—what we call the resupply van— won't arrive for another week, and by then the mission will be over. To conserve as much water as possible, we've implemented a staged cleaning station for dirty dishes with three buckets: first, we scrape off the dirty dishes, then soak them in the second bucket, then scrub them with soap in the third, and finally dry them on the rack.

No showers. No toilet flushing. No brushing our teeth with water.

No water on Mars.

In bed that night, I curl into my black sleeping bag and tug the liner up to my chin. Through paper-thin walls I hear Nils typing on his keyboard at his workstation and Michaela and Britaney howling with laughter in the kitchen below my bed. The LED adjustable lamp glints through the dark at its lowest setting, casting the compact room in a faux twilight.

In these moments of quiet, when my body ceases moving and my mind is free to run wild, anxiety slides into my body with the ease of a familiar guest. Quickly, so quickly, it hijacks my body schema—as it has done throughout my life, and the lives of those generations before me.

It does not matter that nothing is wrong. It does not matter if, moments ago, waves of happiness or gusts of sadness crested into me.

Anxiety doesn't care about reality.

This time, the sudden awareness that I am in the middle of nowhere burrows into me. Everything we have at the Hab is everything we need, and yet the understanding tugs at my mind that if something goes wrong, we are trapped.

Whether CAPCOM is 100 million miles away—if we were truly on Mars—or hundreds of miles, as it is from this simulation, back on the mainland of the United States—either way, we are alone.

But my heart rate remains steady, and when I exhale, I let out my tension, picturing it filling the room, a noxious gas that seeps through the airlock and dissipates into the Martian atmosphere.

I trained for this.

Not even one year ago, I was nestled into a brown leather recliner recovering from my third—and final—breast-cancer risk-reduction surgery, fat grafting, to smooth out the look of my breast reconstruction. For days I could barely stand, for weeks I could hardly walk. For months I was terrified of something going wrong. An implant flipping; a residual piece of scar tissue restricting my movement; infection and sepsis; falling and an implant popping, then I'm back in the hospital, undergoing yet another surgery.

Hours—entire days—of strength training, of teaching my body how to work again. Regaining mobility after my chest was one big mound of scar tissue, my pecs so tight I cried when they stretched upward.

It was during one of those hours in the gym, learning to master my new body, that I decided to apply to NASA to become an astronaut. Months later, when the terse rejection email arrived, a small part of me wanted to shut my eyes, but a louder, more insistent part, perhaps the shadow of my dad perched on my left shoulder, or the specter of Grannell embedded in my mind, reminded me that this is just a learning experience. Just the beginning.

That was when I learned about HI-SEAS, the Hawai'i Space Exploration Analog and Simulation, a research station for analog Martian and Lunar missions. Run by the International Moon-Base Alliance, the HI-SEAS habitat has been home to half a dozen NASA and DARPA missions.

If I can't go to space yet, I told myself, this is a good next step.

Three months after I submitted my application materials, the director of HI-SEAS and my now-commander, Dr. Michaela Musilova, surprised me with the email that I was accepted to both a Lunar and Martian analog.

"Your choice," she said. I couldn't believe my luck.

Three months before the start of the mission, I began training in earnest. Hundreds of hours devoted to running in a hypoxic high-altitude room, learning how to function through severe oxygen deprivation. A setting that mirrored the elevation of the Hab, 2,500 meters above sea level—and spaceflight.

Pushing my body to new heights that towered wildly higher than any I'd dreamt of before my mastectomy.

Dozens of hours spent diving deep below the ocean surface, learning how to breathe regularly with gear covering every centimeter of my body. Gear not unlike my EVA suit: a full six-millimeter neoprene wetsuit, gloves, fins, mask, tank, buoyancy-control device, regulator, dive weights, depth gauge, compass—so much gear that I had to waddle into the water until it could pull me under the surface.

Learning how to handle neutral buoyancy in extreme environments, just as astronauts do on the International Space Station and while they train underwater in NASA Johnson's Neutral Buoyancy Lab.

Breathe in through my mouth, out through my mouth. Hear the faint whoosh of exhalation as my CO_2 spittle flies into the regulator. Whatever is happening above or below does not—cannot—matter; all that matters is that I keep breathing, keep moving.

Learning how to manage unfamiliar, life-saving gear. Get something wrong when I don my gear, and I die. Miscalculate how much oxygen I have in the tank, and I die. Double-check, triple-check my own and my partner's gear—nobody else is going to save us.

But nothing—not even neutral buoyancy—compared to the feeling of weightlessness. My training highlight, the time I felt closest to being a real astronaut: flying in zero gravity, when my body moved of its own accord, independently of my mind. A specially modified Boeing 727 flew in parabolic arcs, achieving twenty to thirty seconds of zero g each time the plane climbed to the top of the parabola. The sensation of zero pressure, no weight, was utterly alien, and no matter how hard I tried, I learned it was impossible to direct my movement.

Flujo. Going with the flow.

Up there, floating weightless 30,000 feet above the Earth, I learned to surrender any notion of control.

And in doing so, for those few moments of weightlessness, I was free.

From my bed in the Hab, on simulated Mars, as my eyes shut and the Sol ebbs away, I imagine a foreign sky of stars winking into existence way above the dust storm. As my breath steadies and I feel myself drift away, the thought floats into my head that my body and my mind have accomplished extraordinary things.

Who's to say I can't handle Mars?

Outside, the dust storm is still wreaking havoc when I wake up. Groaning, knowing this likely kills our chances of a morning EVA to conduct research, I roll over on my side, fumble for my glasses, and trudge to the bathroom across the hall.

Pop in a tablet of toothpaste, sprinkle one single drop of water on my toothbrush, brush until the dry chalk taste is so unbearable that I have to spit it out. Take a wet wipe and rub my hands, unscrew my contact-lens carrier and poke one into each eye, praying the dirt still smudged on my fingers doesn't make its way into my eyes. Twice, I run coarse bristles through my hair, trying to wrestle them through the growing knots. Spray the roots with dry shampoo, rub lotion on my face, draw on liquid eyeliner and groan at the black smudges now along my eye crease.

When I make my way downstairs, Nils is typing away at his computer in a flight suit, baseball cap, and over-the-ear noise-canceling headphones wrapped around his head. We nod good morning to each other.

A wooden spoon clangs against steel in the kitchen, and when I turn around the bend, I find Elisha already standing over the stove, stirring crumbs of freeze-dried raspberries into a pot of oatmeal.

"Still storming?" I ask her.

She nods. "Seems to be moving away from us, so I bet it'll lift by this evening. But Michaela says no daytime EVA."

From the circular window, I can see nothing but fog and diffuse morning light.

"Dammit," I groan.

"I know," she says, clanging the spoon against the pot once more and setting it aside. "But she said if it clears by sunset, we can do a nighttime EVA."

I blink. "Really?"

She looks up from the steaming pot and grins.

"Yeah. Get ready for our first nighttime spacewalk," she says, and I smile.

Oatmeal is the de facto breakfast of choice, same as our last nine Sols. Wearily, I rub my eyes and pick up my spoon, dipping it into the dry sludge. No matter how much water I want to conserve, I vow to myself, after the mission I'll never eat oatmeal again.

The five of us sit along the white picnic table, and while I spoon the bland oats into my mouth, trying to ignore the taste, Michaela begins our morning briefing.

"We're unable to conduct a daytime EVA for obvious reasons," she says, waving vaguely toward the window. "Britaney, can you submit a request to CAPCOM for a nighttime EVA?"

Excitement runs like an electric current through the crew, wiping away our disappointment.

"Yes, definitely," says Britaney.

"What is a nighttime EVA like?" asks Elisha.

"You'll see," Michaela says, winking. "We'll be limited to the area directly around the Hab, but we'll still have an unobstructed view of the night sky."

My breath catches. The night sky of a place so isolated from humans, so far from light pollution, that it resembles Mars.

Excitement thrumming through me, the taste of the oatmeal long forgotten, I am surprised when the spoon rising to my lips is empty.

"The rest of you," Michaela says, "catch up on whatever work you can do without an EVA. Elisha and Sarafina, we can just do leftovers for lunch, so all we'll need you to prepare is dinner."

"Sure," we say.

"Great," she says. "Then let's hope this dust storm passes and we can get outside tonight."

Throughout the rest of the day, I work through three derivations, type emails to my advisor, take notes on academic research papers.

For an instant, in the middle of struggling through the second paper I'd planned to work through today, anxiety overwhelms me in one enormous spell. The author is talking about the radioactive decay of elements in a supernova explosion, and I can't wrap my mind around the physics.

I blink away burning tears, wishing I was upstairs in bed and not in the middle of the downstairs workspace, and the part of me that is steady and kind tries tugging at the sickness poisoning my thoughts. I feel the internal conflict as anxiety in my body; it lights coals in my guts and chokes off my throat.

But the small, steady part is growing stronger, more resilient, with each anxiety cycle it manages to get through.

It begins as a whisper: *You're okay if you're not as productive as you think you should be.*

You're exactly where you need to be, right here, in this moment.

And then, that small voice gets stronger.

Your self-worth is not tied to your output.
You are worthy just being you.
You are you, and that is enough.

My breathing calms and the buzzing in my head begins to subside, and with each passing moment the sounds of the Hab get louder. The clickety-clack of keyboards around me. Nils sipping from the straw of his navy-blue water bottle. The tearing of construction paper as Britaney cuts through a ream of yellow-and-pink pages. Elisha wearily sighing as she stares at lines of code. Michaela giggling as she reads something on her desktop. The faint trickle of water hydrating the hydroponic lettuce farmstand.

I heave another giant exhale.

I am okay.

Not every anxiety cycle is quicker than the last. Sometimes, I feel like I'm falling backward, and I stumble upstairs to get my emergency medication, tucked into a gray-and-white zippered bag with the rest of my belongings.

But with every cycle, my thought patterns are becoming clearer. I am beginning to understand how they fit together: first, the self-loathing; then, the inward criticism; finally, deeply ingrained guilt and shame that swirl around in a chaotic gust at the end of each cycle. I'm noticing just how vicious I can be to myself.

I'm accepting that this is where I am right now.

And with each cycle, my anxiety eases by a fraction of a degree.

After a quick lunch of leftover beans and quinoa—Michaela eats the last bits of Space Noodles to spare us—we undergo our mandatory one-hour-per-day exercise. Britaney cycles on the bike, Elisha practices yoga on a blue mat left over from a previous mission, Nils sprints on the treadmill, Michaela does handstands. In between my lunges, I spot Britaney monitoring our levels on the big screen and eventually propping open the door to the cleaning supplies.

"Too much CO_2," she says.

Since we don't have showers, after my workout I run upstairs to clean my body with wet wipes. I ran out of fresh clothes three Sols ago, so I sniff my pile of used clothes and don the least smelly ones. Grimacing at the faint scent of sweat now embedded in my skin, I trudge back downstairs and check the clock.

Three o'clock. Still one hour before dinner prep and two hours before the CAPCOM communication window opens.

On the floor below the window is a long red cushion with vertical white stripes. I grab my iPad and slide onto it, tucking my knees into my chest, and gaze out at the walls of the Hab. On each square inch of white space around me are multicolored stickers of space agencies, handwritten messages from previous crews, a signed poster of Captain Kirk, and NASA exoplanet posters with watercolor skies and barren, rocky surfaces.

Gazing blankly at the decor, I tug my knees closer. On my iPad, I click on another supernova paper and pick up the white stylus to begin highlighting. Five minutes pass, and I'm staring at the same sentence not one-quarter of the way through the introduction section. None of the words have registered in my brain.

These are the hardest hours in the Hab, when all of my tasks are done and my brain is fried and I have nothing to do but watch the clock tick. There are no social-media distractions, no television, and no texting with friends.

Instead, time plods forward, as though Mars is orbiting the sun in a time warp and everything moves more slowly here.

Slowly, I place the stylus against the side of the iPad, letting it magnetically fall into place.

I could keep trying. I could scramble back to my desk, pull the paper up on a wide-screen monitor and force myself through it. I could wrestle with the derivations, bang my head against the wall.

Or I could rest.

Letting my head fall into the cushion I've scooted into the wall of the Hab, I shut my eyes and remember the ten thousand hours I spent sweating on the tennis court, when tears lined my eyes and I forced my body to continue past the pain, tune it all out and push-push-push.

I remember downing those midnight coffees alone in my house in Austin as I blearily rubbed away the sleep from my eyes and pushed through physics calculations, to get that A+ so maybe I would be accepted, loved.

From the millstone of my memory, a vision of ten-year-old me trudges from one doctor's appointment to the next, complaining of the phantom pain in my stomach.

I remember my mom and dad arguing at a gas station and watching from that black leather backseat, vowing to myself that I will make things better if I just work hard enough.

What if, instead, I choose rest?

From a corner of my mind, Grannell rises from her favorite crimson butterfly chair, tugs me into a hug, and whispers in my ear. *Make good choices.*

I am learning, here on Mars, that choosing rest is good.

It's not a weakness. It's strength.

Still, even as I lean into rest, guilt gusts into my chest and threatens to wind me into a panic. But in the back of my mind, I remind myself to listen to my body, to notice my foggy thoughts and tired eyes.

To let them be.

I spend an hour on that cushion hovering in and out of sleep, luxuriating in the calm before our schedule ratchets up again. When the clock strikes five, Elisha calls to me that it's time to prepare dinner, and together we boil the water, knead the dough, bake the garlic knots, season the tops with garlic and rosemary flakes.

By 5:30, shafts of light are spilling through the window, and we heave a collective sigh of relief. All five of us pause our tasks to plug in our devices, and Britaney switches the Hab back to normal power mode.

I cook the last of our carrots while Elisha boils the spinach, and when we gather at the table, we swallow the food in great gulps. Every few seconds, one of us glances outside the window.

By six, Michaela stands and strides to the window for final confirmation. Whirling around, she smiles.

"CAPCOM approved your request for a night EVA. Be sure to send in your mission reports while the comms window is still open, and then get dressed," she says. "We're going outside."

When we exit the airlock, the cold is what hits me first. Even with the EVA suit, life support, and two layers of socks, night on Mars is brisk and unforgiving.

The sun dips just below the horizon, dust storm long gone, no traces of clouds remaining. Stretching before me is an infinite amount of barren land, utterly empty and so enormous that my mind snags on it, stunned. How can I see so much, yet so little?

And then I look up.

All above me, the sky is already aflame with dangling stars winking into existence. Ten, twenty, one hundred, more stars than I have ever seen, already so bright they cast the Martian surface in a transcendent glow. From above the Hab, their light glitters against the white, flickering halolike against the dome.

My breath catches. I have never seen anything so beautiful.

The sky is an inky blue, and the crew lines up underneath the Martian sky to watch it turn from dusk to black. All of us are silent.

We remain like that for minutes or hours, none of us saying anything, frozen in that Martian time warp that has ensnared us all. Every time a new star winks into existence, I'm struck anew by the utter vastness, the scale of land and sky and universe unspooling around me.

Up there, I imagine, somewhere between those stars, is where each of us goes at the end of it all. Our ashes and atoms sprinkle through the stars, the oxygen in our blood and the calcium in our teeth enriching the chemical makeup of the cosmos until one day another life sprouts somewhere else.

Grannell is both up there yet also in my mind, my dad is on my shoulder and in Mexico, my mom is in my heart and in Texas, I am here and I am up there.

For an instant, time is spinning up and rewinding, and all of a sudden, I'm six years old, curled up on my side in bed, Stewy Bear wrapped in my arms, and my dad is kissing me good night.

It spins up again—

"Astronomy isn't for you," says the voice of the astronomer when I am ten years old at science camp.

It whirls forward, and I am twelve.

"You can achieve," my dad says to me, *"whatever you set your mind to."*

Images slide by, one after the next, and I am back in Mikan's observatory, staring up at his well-worn blackboard, and he is saying that a scientist is someone with a burning *curiosity* for the world. The world tips forward, and from all around me comes Dr.

Wheeler's gravelly voice: *"You will make an excellent scientist, Sarafina."*

I am lying underneath the stars in West Texas, I am crumpled up on Ahmad's floor, I am onstage delivering the commencement speech, I am walking through the double-paned doors at UC Berkeley, I am lying in a hospital bed, and I am touching my new body.

Taylor is wrapping me into his arms and I am breathing in Comet's fur. From a crease of my mind, I hear Kyoko, and her words settle into me.

"You're exploring the universe out there to explore the universe within yourself."

Every scene is melting into the next, and it is impossible to separate myself from it all, I am all of those moments, yet they are not all of me—I am *more*.

Amongst the waves soaring and crashing around me, I am seeing it all unfold, every moment of my life turn by turn, but inside me, for the first time since I was a child, maybe for the first time in my life, I feel as though I have found myself, like I have finally settled onto the ground in my soul.

"One day," my dad tells me when I am twenty, *"I'm going to see you up there."* And he points up into the stars, all the way to Mars.

ACKNOWLEDGMENTS

Memoir is difficult. There are parts of my life that were so extraordinarily painful to relive that I didn't want to keep going. But there are also moments of incandescent, wild joy. I hope both resonated with you.

Thank you, reader, for picking this book up and for giving it a chance. I wrote *Starstruck* as a form of catharsis. Healing is a messy, painful thing. Each of the individuals listed here aren't just integral to this book—they are integral to me, as a human being. I also wrote this as a form of liberation—for myself and for others. I hope that in reading this, you feel less alone.

My deepest love and gratitude go to:

Taylor, you are my everything. Thank you for being my rock, my wings, my best friend. Thank you for bearing with my (recovering) workaholism and supporting my absolutely insane decision to write a book while pursuing a PhD. For the many, many dog walks and hugs, beverage refills and chocolate bars, pep talks and patience, but especially the unconditional love that you give so freely. I love you.

To Comet, the best, most anxious dog I've ever known. This book exists because Comet knows how to lick my face when I'm sad, curl into me when I'm anxious, and remind me that sometimes all one needs is a walk outside and some treats. Thank you for the cuddles. You are my baby forever.

To my literary agent, Melissa Danaczko: I love you. This book would not exist without you (no, really, it wouldn't). From those earliest stages of fleshing out what this book means, to guiding me through the spectacularly opaque world of publishing, to hopping on countless calls and easing my nerves and guiding me through anxiety attacks—I literally don't know how you do it. You approach everything with such compassion and care. You have become so much more than my agent; you have become a dear friend. You have taught me so much about writing but even more about life. Thank you for taking a chance on me. Thank you for being you. Thanks also to everyone at Stuart Krichevsky Literary Agency, including Stuart Krichevsky and Hannah Schwartz. What an absolutely all-star team!

To my remarkable editor, Jill Schwartzman, for believing in the power of my story. This book is immeasurably better thanks to your enthusiasm, open heart, wisdom, guidance, and (lightning fast!) edits. It really is a dream come true to work with you.

To Charlotte Peters and Patricia Clark, for your insightful comments and tireless support in navigating some of the most daunting aspects of the publishing process. Maya Ziv for introducing me to Melissa and always being the biggest cheerleader! The entire team at Dutton, for your unrelenting support and extraordinary efforts to bring this book into the world. It's an honor to work with each of you!

To my dad. Well, I once wrote fifty things I love about you— now here's a whole book! No words in any language can encapsulate how deeply I love you and how grateful I am that you're my dad. Thank you for being my best friend.

To my mom. I know that our relationship is complicated and beautiful—still, thank you for loving me as I love you. Thank you for giving me your name. Thank you for taking me to our heartland, Egypt, and for understanding that identity is shaped by culture, by the intangible, invisible threads that make up our DNA. Thank you for your strength and your strong sense of intuition. Thank you for taking care of me and always being my cheerleader. I love you.

To the badass women who reviewed the science sections of *Starstruck*, Hannalore Gerling-Dunsmore, Dreia Carrillo, Logan Pearce, Jackie Champagne, Taylor Hutchison. This book is far better thanks to your rigorous review and thoughtful edits. Thank you to Micaela Bagley and Rebecca Larson, who have helped me over more astronomy hurdles than I can count. The entire group of WISDOM (Women in STEM Doing Our Mightiest), thank you for your support and your friendship.

To my therapists over the years: Shina Lee, Kyoko Tsuchiya, Cameron Kemper, Sara Champie—thank you for keeping me safe. Thank you for guiding me on my lifelong journey into myself. I'm so honored to have you all as mentors.

To my friends who have read countless excerpts, especially Sus Lodge-Rigel, who was with this book from birth to the bittersweet end. Thank you for reminding me that love and community are the universe's greatest gifts.

To the women who inspire me, especially Dr. Anne Peled for saving my life and being my real-life superhero, and the Breasties (the worst club, best members!). We are stronger than we know.

To my teachers and tutors, especially Dr. Peter Nugent, Dr. J. Craig Wheeler, Frank Mikan, and Scott Gerlach, for igniting my curiosity, advocating for me, and believing in me.

To the science communicators who strive to share the magic of their fields with others. It's not easy but the world is immeasurably better for it—keep fighting the good fight.

To the authors who inspire me—I'm a fantasy nerd who hardly ever reads nonfiction. This was an exercise in learning to write memoir as much as it was endeavoring to understand the complex threads that make up a life. Thank you to the authors who build fantastical worlds to which we can escape, especially when life gets hard. You have gotten me through more difficult times than I can count.

To any person who has experienced domestic violence and abuse, I am so sorry. Healing is a lifelong process but please remember that you are not alone—and this isn't your fault. You are worthy just by being you.

And finally, to all the young girls who dream of the stars. Don't listen to the naysayers, to those who question your abilities. You got this. We need you.

ABOUT THE AUTHOR

Sarafina El-Badry Nance is an Egyptian-American astrophysicist, analog astronaut, author, science communicator, *Sports Illustrated* swimsuit model, and fervent women's health advocate. Passionate about increasing science literacy, she uses a variety of mediums (social media, books, academic publications, film, and TV) to share the magic of the universe with the world. After being diagnosed with the cancer-causing BRCA mutation and having a preventative double mastectomy at age twenty-six, she publicly advocates for genetic testing, self-checks, and equity in healthcare. For her work, Sarafina has been named one of *Forbes'* "30 Inspirational Women" and was on *Forbes'* list of "30 Under 30" and the Arab America Foundation's "40 Under 40." She lives in Berkeley with her partner, Taylor, and her dog, Comet.